Statistics and Computing

Springer
New York
Berlin
Heidelberg
Barcelona
Budapest
Hong Kong
London
Milan
Paris
Santa Clara
Singapore
Tokyo

Statistics and Computing

Härdle/Klinke/Turlach: XploRe: An Interactive Statistical Computing Environment
Venables/Ripley: Modern Applied Statistics with S-Plus
Ó Ruanaidh/Fitzgerald: Numerical Bayesian Methods Applied to Signal Processing

Joseph J.K. Ó Ruanaidh
William J. Fitzgerald

Numerical Bayesian Methods Applied to Signal Processing

With 118 Illustrations

Springer

Joseph J.K. Ó Ruanaidh
William J. Fitzgerald
Department of Engineering
University of Cambridge
Trumpington Street
Cambridge CB2 1PZ
United Kingdom

Library of Congress Cataloging-in-Publication Data
Ó Ruanaidh, Joseph J. K.
Numerical Bayesian methods applied to signal processing / Joseph J.K. Ó Ruanaidh, William J. Fitzgerald.
p. cm.
Includes bibliographical references and index.
1. Signal processing—Statistical methods. 2. Bayesian statistical decision theory. I. Fitzgerald, William J. II. Title.
TK5102.9.078 1996
621.382′2′0151954—dc20 95-44635

Printed on acid-free paper.

Production managed by Frank Ganz; manufacturing supervised by Joe Quatela.
Photocomposed pages prepared from the author's LATEX files.
Printed and bound by Edwards Brothers, Inc., Ann Arbor, MI.
Printed in the United States of America.

9 8 7 6 5 4 3 2 1

ISBN 0-387-94629-2 Springer-Verlag New York Berlin Heidelberg SPIN 10522339

Dár muintir

To our families

Acknowledgments

"Ar scáth a chéile a mhaireann na daoine"

Thanks are due to Jebu Rajan, Robin Morris, Simon Godsill, Miao Dan Wu, Jacek Noga, Teng Joon Lim and Anil Kokaram for proofreading sections of this book. We would also like to thank the network team – Anil Kokaram, Pete Wilson, Ray Auchterlounie and Ian Calderbank for doing such a sterling job in managing the lab Novell network and Alex "The LaTeX Baron" Stark for maintaining the document preparation software so well.

We also wish to acknowledge Radford Neal, Sibusiso Sibisi, Simon Godsill, Thomas Reiss, William Graham, Anthony Quinn and David MacKay for their help and advice.

We are delighted to thank the Engineering Department, University of Cambridge for providing such a wonderful working environment where all of this work was carried out. We would like to thank the Department of Electronic and Electrical Engineering, Trinity College, University of Dublin for the generous use of facilities during a critical stage in the final phases of the book's preparation.

Joseph J.K. Ó Ruanaidh
William J. Fitzgerald
Cambridge 1995

Glossary

AIC	Akaike's Information Criterion
AR	Autoregressive
ARMA	Autoregressive Moving Average
BFGS	Broyden Fletcher Goldfarb Shanno
cdf	Cumulative Distribution Function
CG	Condensed Gibbs Sampler
cpdf	Conditional Probability Density Function
DFP	Davidon Fletcher Powell
DVM	Dummy Variable Method
EM	Expectation Maximisation
FG	Full Gibbs Sampler
HMC	Hybrid Monte Carlo
GA	Genetic Algorithm
GPL	General Piecewise Linear
i.i.d.	Independent Identically Distributed
LS	Least Squares
MA	Moving Average
MAP	Maximum a Posteriori
MCMC	Markov Chain Monte Carlo (MC^2)
MCMCMC	Markov Chain Monte Carlo Model Comparison(MC^3)
MDL	Minimum Description Length
ML	Maximum Likelihood
PC	Programmable Computer
SIR	Sampling Importance Resampling
SNR	Signal to Noise Ratio
pdf	Probability Density Function
VFSR	Very Fast Simulated Reannealing
VM	Variable Metric

Notation

The following notational conventions are used in the main text:

a	scalar
$\mathbf{b}$	column vector
$\mathbf{b}^{\mathrm{T}}$	transpose of $\mathbf{b}$
b_i	i^{th} element of $\mathbf{b}$
$\mathbf{R}$	matrix
$\mathbf{R}^{-1}$	inverse of matrix $\mathbf{R}$
$\lvert\mathbf{R}\rvert$	determinant of matrix $\mathbf{R}$
R_{ij}	j^{th} element of the ith row of $\mathbf{R}$
$\mathbf{I}$	identity matrix
$\mathrm{E}(.)$	expectation operator
$p(\mathbf{x})$	joint pdf for the elements of $\mathbf{x}$
$p(\mathbf{x}, \mathbf{y})$	joint pdf for the elements of $\mathbf{x}$ and the elements of $\mathbf{y}$
$p(\mathbf{x} \mid \mathbf{y})$	joint cpdf for the elements of $\mathbf{x}$ given the elements of $\mathbf{y}$
$P(A)$	probability of event A
I	prior information
N	number of data points
$N!$	$N!{=}N.(N-1).(N-2)\cdots 3.2.1$ where N is an integer
∇	gradient operator
$\mathcal{O}(.)$	order of approximation
$\lvert x\rvert$	x scalar; absolute value of x
$\Re^M$	M dimensional space of real numbers
$f(\mathbf{x})\rvert_{\hat{\mathbf{x}}}$	function $f(\mathbf{x})$ evaluated at the supremum $\hat{\mathbf{x}}$
$(0,1]$	real interval from 0 to 1, including 1 but excluding 0
$x \leftarrow f(x)$	x is random number drawn from pdf $f(x)$

Contents

1
Introduction

This book is concerned with the processing of signals that have been sampled and digitized. The fundamental theory behind Digital Signal Processing has been in existence for decades and has extensive applications to the fields of speech and data communications, biomedical engineering, acoustics, sonar, radar, seismology, oil exploration, instrumentation and audio signal processing to name but a few [87].

The term "Digital Signal Processing", in its broadest sense, could apply to any operation carried out on a finite set of measurements for whatever purpose. A book on signal processing would usually contain detailed descriptions of the standard mathematical machinery often used to describe signals. It would also motivate an approach to real world problems based on concepts and results developed in linear systems theory, that make use of some rather interesting properties of the time and frequency domain representations of signals. While this book assumes some familiarity with traditional methods the emphasis is altogether quite different. The aim is to describe general methods for carrying out *optimal signal processing.*

In this book the treatment given to digital signals may be divided into two main areas:

1. The first main area is *signal enhancement*, where one may wish to modify a signal in some way to improve its quality. For instance, digital filtering is often used as a means of noise reduction to improve the subjective quality of audio signals. In this book, we shall describe several methods for the removal of clicks from digital audio signals.

2. The second area, and one that is of great interest to us, is *data analysis*. Assuming the signal is carrying information relating to some physical phenomenon, our aim is to extract this information using some suitable means. One example from physical chemistry is molecular spectroscopy, where there is considerable interest in the constituent frequencies of the signal obtained when a sample of a substance is made to resonate by an externally applied magnetic field. Different kinds of chemical bond are characterized by distinct groups of frequencies. These signals contain important information that may lead to a determination of the molecular structure, and hence explain known properties of the substance, as well as yield predictions of hitherto unknown properties.

Very often, explicit mathematical models exist for the form of the signal and of the noise. The signal may be the result of some known physical property inherent in a system. In these cases the behaviour of the signal is characterized by the signal parameters whose values are usually unknown. Sometimes there can be many possibilities for the signal model. The aim is to find the most appropriate model to describe the data as well as to estimate the parameters. The random nature of noise can make it impossible to determine what exactly is occurring so this is easier said than done.

All signals are corrupted by noise. Noise may be due to any number of different causes, such as damage to a storage medium, electrical interference, material impurities and temperature etc. Since noise is a fact of life, the basic question arises as to how best to deal with it.

The task of parameter estimation and model selection is an exercise in reasoning in the presence of uncertainty. The toolbox that we use to solve such problems has been in existence for hundreds of years. Bayesian methods operate by representing degrees of belief in opposing hypotheses (e.g. the value of a parameter is x rather than y) in the form of probability densities. This approach has great intuitive appeal, but has traditionally met with stiff resistance from the statistical establishment who subscribe to the theory that probabilities are the relative frequencies of occurrence of repeated events.

It was comparatively recently that the Bayesian framework was placed on a sound theoretical footing. The formal rules that are used to manipulate probability densities in a consistent fashion were mathematically established by Cox [24] in 1946. These rules arise from a set of conditions (or "qualitative desiderata") for a measure of probability:

1. *Degrees of plausibility may be represented by real numbers.*

2. *The system must be internally consistent and be consistent with simple rules of common sense.*

These desiderata apply only to the plausibility of propositions and make no reference to frequencies, random variables or ensembles. Remarkably, these two desiderata are sufficient to uniquely define the form of Bayesian probability theory.

Bayesian probability theory is not new, and has been in use in one form or another since the time of Laplace (who used it to calculate the probability that the mass of the planet Saturn lay within certain bounds [57, 58]). Recently however, there has been a massive growth of interest in Bayesian methods. There have been some great successes such as Bretthorst's application of the method to NMR spectroscopy [13, 14, 15, 16] as well as more general applications which require the resolution of closely spaced parameters [18, 20, 68].

There is nothing magical in the success of Bayesian techniques. As noted by Jaynes [57], in data analysis often the wrong question is asked, and even in those cases where the right question is asked, it often gets answered in the wrong way! Simple examples include the use of the discrete Fourier Transform power spectral density for spectral analysis, which as Bretthorst shows [13], is perfectly adequate for determining widely spaced frequencies, but is totally ineffective at resolving closely spaced frequencies. Another (infamous!) example is the use of smoothing filters to suppress noise in data sets [22], which usually have the effect of introducing artifacts into the data [57]. However, it must be added that such approaches are usually motivated by the relative lack of computation involved rather than gross ignorance on the part of the analyst.

This brings us to the core of this research. Recent developments in computer technology, to a large extent, have made Bayesian methods feasible for use in many real world problems with large numbers of parameters and with complex signal models. At the same time, the development of new techniques for the optimization, integration and random simulation of probability densities has provided the mathematical tools required to implement the Bayesian paradigm.

There are three different aspects to numerical Bayesian inference:

Optimization: The most simple objective is to locate the peak of the probability density in order to find the most probable fit to a data set. In general this requires non-linear multidimensional optimization.

Integration: The computation of *model evidence* (for model selection) and of *marginal densities* (for parameter estimation) requires the evaluation of multidimensional integrals. In general, except for some very simple examples, integration cannot be carried out analytically. The development of efficient strategies for the computation of Bayesian integrals is therefore a fundamental prerequisite to the general application of Bayesian methods.

Simulation: The simulation of a probability density, or more precisely, drawing typical random vectors from a probability density, is also of interest in some applications.

A recurring theme in this work is the close relationship that exists between the apparently disparate areas of optimization, integration and random simulation.

The plan of this book is to begin with a review chapter on Bayesian inference in signal processing, and to demonstrate the main principles by solving analytically tractable problems. This is followed by a section consisting of two chapters on numerical methods for the analysis of previously intractable problems. The third section applies these techniques to real and simulated data.

The scope of the applications in this book is very broad and includes the replacement of missing samples in digital audio recordings and the detection of abrupt discontinuities in oil well data, as well as some of the more traditional applications of Bayesian methods to data analysis.

The following is a detailed outline of the remaining chapters of this book:

Chapter 2 discusses the main principles of Bayesian inference applied to signal processing. Topics discussed include prior probabilities, marginalization for the removal of nuisance parameters, and the evidence framework for the comparison of competing models. The general linear model is introduced and the principles of Bayesian inference are illustrated using simple examples.

Chapter 3 discusses numerical approaches to the optimization, integration and simulation of probability densities. The chapter begins with a presentation of the principle of asymptotic normality. Optimization methods appropriate for use in Bayesian signal processing are then discussed. These include local gradient based algorithms such as conjugate gradients and variable metric, as well as global algorithms such as simulated annealing and genetic algorithms. We then discuss the integration of multivariate probability densities using Gaussian quadrature, asymptotic approximations and Monte Carlo methods. The chapter concludes by focusing on Monte Carlo methods in particular, presenting algorithms specific to computing evidence and marginal densities.

Chapter 4 is a review of Markov chain Monte Carlo methods. The canonical distribution for the energy of states is discussed and from it the thermodynamic relationship between energy and probability density is derived. The Gibbs sampler, the Metropolis-Hastings algorithm and the dynamical Hybrid Monte Carlo method are introduced as methods for sampling difficult probability densities. The theory of optimization using simulated annealing is presented. The determination of model evidence using thermodynamic free energy methods is introduced.

Chapter 5 derives Bayesian methods for the retrospective detection of abrupt changes in data using the Bayesian piecewise linear model.

Examples investigated include the detection of a simple change of mean in a Gaussian process, as well as the detection of a change of coefficients in an autoregressive random process. Algorithms for the recursive update of probability densities are developed. The more difficult problem of detecting multiple changepoints in non-Gaussian noise environments is then addressed. The method of simulated annealing using the Gibbs sampler is employed to solve this problem.

Chapter 6 is concerned with the interesting problem of restoring missing samples in digital audio signals. The signal in question is modelled as a locally stationary autoregressive process and the missing samples are restored by interpolation. Three separate solutions to the problem are developed. Two of the approaches are standard to this kind of missing data problem, namely maximum likelihood and the EM algorithm. The third approach discussed is to use a random sample interpolant produced using the Gibbs sampler. The performance of the Gibbs sampler is compared with, and shown to be superior to, that of EM and ML for the restoration of both real and synthetic data.

Chapter 7 applies the numerical methods described in chapter 3 and chapter 4 to marginalization and model selection for a number of different data models. The initial examples are analytically tractable so the numerical results can be compared with known theoretical results. Finally we discuss examples where the integrals may not be computed in closed form. In this chapter a particularly difficult problem is also attempted, namely the determination of the decay rates of the superposition of exponential curves in noisy data.

Chapter 8 concludes the book with a summary of the main results and suggests directions for future research.

2

Probabilistic Inference in Signal Processing

In this chapter, the fundamental concepts and techniques underlying Bayesian inference are reviewed. We begin with a definition of the key problem of data analysis, which is to interpret data in the presence of noise. We advocate a Bayesian approach to this problem.

The likelihood function for a data model and common prior probabilities are defined. Bayesian parameter estimation and Bayesian model selection are illustrated by simple examples.

A compact matrix formulation of a very common class of signal model, known as the general linear model, is presented. The chapter concludes with an analysis of synthetic data using the general linear model to illustrate Bayesian techniques.

2.1 Introduction

We are concerned with two basic problems in data analysis. First, we wish to choose from a set of candidate models that which is best supported by the data, a task known as model selection. Second, we wish to determine the values of unknown parameters from observed data sets; a task loosely termed parameter estimation.

In all cases a finite number of observations is presented. The data is assumed to consist of the sum of two components: a time dependent systematic component and a random process.

The $N \times 1$ data vector $\mathbf{d}$ is assumed to be related to the time vector $\mathbf{t}$ by:

$$d_i = f(t_i, \theta) + \mathbf{e_i} \tag{2.1}$$

where θ is the vector containing the parameters of the signal model. The residual term e_i is assumed to be sampled from a known random process.

The choice of signal model for $f(t_i, \theta)$ depends on the specific application. One example that is very thoroughly investigated by Bretthorst [13] is the single frequency model:

$$f(t_i, \theta) = A \cos \omega t_i + B \sin \omega t_i \tag{2.2}$$

In this signal model, the parameter vector θ has three components: phase amplitude A, quadrature amplitude B and angular frequency ω. This signal model is of special importance in Bayesian spectral analysis.

The assumption of a noise model corresponds to selecting the statistics of the random component of the data. The most commonly used noise model is the Gaussian model:

$$p(e_i) = \frac{1}{\sqrt{2\pi\sigma^2}} \exp\left[-\frac{{e_i}^2}{2\sigma^2}\right] \tag{2.3}$$

This noise model leads to a tractable analysis in many cases. Theoretically it can also be considered to be a general noise model because of its conservative nature. Three justifications for its adoption in the absence of any information regarding the noise statistics include the central limit theorem [89], Herschel's theorem [13] as well as the principle of maximum entropy [17, 57].

2.2 The likelihood function

A proper interpretation of the likelihood function is very important from our point of view. The likelihood is the probability of realizing the data given the value of the parameters, the signal model and the noise statistics. Denoting the specific choice of signal model and noise statistics as I_k one may write:

$$p\left(\mathbf{d} \mid \mathbf{\Phi}, \mathbf{I_k}\right) = L(\mathbf{\Phi}\,;\,\mathbf{d}) \tag{2.4}$$

where the vector $\mathbf{\Phi}$ includes both the parameters of the signal model $f(t\,;\,\theta)$ and the noise parameters.

The data is assumed to consist of the signal with added independent identically distributed (i.i.d.) noise. In this case, the likelihood is of a particularly simple form. It is easy to show [99] that the likelihood function is identical to the joint density of the residuals:

$$L(\mathbf{\Phi}\,;\,\mathbf{d}) = p(\mathbf{e}) \tag{2.5}$$

If the components in the residual vector $\mathbf{e}$ in equation 2.1 are independent and identically distributed then the joint probability of the noise samples is the product of the probabilities of each sample:

$$p(\mathbf{e}) = \prod_{i=1}^{N} p(e_i) = \prod_{i=1}^{N} p(d_i - f(t_i\,;\,\theta)) \tag{2.6}$$

For the Gaussian assumption we can write:

$$\begin{aligned} p(\mathbf{e}) &= \prod_{i=1}^{N} \frac{1}{\sqrt{2\pi\sigma^2}} \exp\left[-\frac{{e_i}^2}{2\sigma^2}\right] && (2.7)\\ &= \left(2\pi\sigma^2\right)^{-N/2} \exp\left[-\frac{\sum_{i=1}^{N} {e_i}^2}{2\sigma^2}\right] && (2.8)\end{aligned}$$

The likelihood function will be of this form for many of the applications in this book.

2.2.1 *Maximum likelihood*

The method of maximum likelihood estimates the parameters $\mathbf{\Phi}$ of the model by locating the supremum of the function $L(\mathbf{\Phi}\,;\,\mathbf{d})$ in parameter space. The rationale behind the method of maximum likelihood is that the parameters and the model are chosen to maximize the probability of observing the data. It is interesting to note that maximizing the Gaussian likelihood of equation 2.8 with respect to the signal parameters is equivalent to minimizing the sum of the squares of the residuals. This gives a theoretical justification for the method of least squares.

Using the method of maximum likelihood to infer the values of the parameters has serious limitations. First, the likelihood does not use information other than the data itself to infer the values of the parameters. No prior knowledge is used about the probable values of the parameters before the data was observed. In some cases a physical understanding of the circumstances surrounding an experiment can suggest that some values of the parameters are impossible. There are cases where maximum likelihood can return parameter estimates outside the sensible range of the parameters.

Second, the likelihood on its own does not limit the number of parameters in a model used to fit the data. In general, the more complex a model the better the fit to the data. For example, a data set consisting of N observations can always be exactly described in terms of a model with N parameters (e.g. a Lagrange polynomial [6, 66]). Suppose a model describes a data set with no error. If the experiment that gave rise to the data set is repeated then, because of the difference in the noise realization, a new model will be needed to describe the new data with no error. Often, the new parameter estimates can be vastly different to the old parameter estimates. With this in mind, Jeffreys [58] argues that model simplicity is the key to maximizing the degree of consistency between parameter estimates computed from independent realizations of the data.

The tendency of the method of maximum likelihood to over-parameterize is well recognized and many attempts have been made to remove this deficiency. Akaike [1, 128] used information theory to determine another approach to parameter estimation using the Akaike Information Criterion (AIC). AIC differs from ML because it incorporates an additional term that increases with the number of parameters. Rissanen [105, 128] and Schwartz [113, 128] independently proposed a similar measure of complexity based on the principle of minimum description length (MDL). Both MDL and AIC operate on the principle that once a model has an adequate number of parameters to describe the data, any increase in the number of parameters does not usually result in a very significant increase in the likelihood. Therefore a term that penalizes complexity that is subtracted from the log likelihood can result in a good compromise between number of parameters and accuracy of fit, and will therefore improve the consistency of parameter estimates.

2.3 Bayesian data analysis

The deficiencies in the likelihood framework are addressed in a very simple and direct way by the Bayesian formalism.

Let us denote the data model as I_k and the parameter set as θ. Also let uncertainties about the values of parameters in a model be represented as probabilities. For our purposes we can write Bayes' theorem as follows:

$$p(\theta \mid \mathbf{d}, I_k) = \frac{p(\theta \mid I_k)\, p(\mathbf{d} \mid \theta, I_k)}{p(\mathbf{d} \mid I_k)} \tag{2.9}$$

where $p(\theta \mid I_k)$ is the prior probability which is chosen to summarize all knowledge of the values of the parameters prior to observing the data. The normalizing factor $p(\mathbf{d} \mid I_k)$ of expression 2.9 is called the Bayesian evidence and is of interest in model selection, as discussed later in section 2.6. The term $p(\theta \mid \mathbf{d}, I_k)$ is the joint posterior density, and it summarizes the state of knowledge about the values of the parameters *after* the data is observed.

It has been noted by many authors [52, 57, 82] that Bayes' theorem as stated in equation 2.9 provides a prescription for describing the learning process , by which prior information is updated in the light of new data. The process is sequential since there is nothing to stop one from using the posterior probability obtained at one stage of an analysis as a prior probability in the next stage of the analysis as more data becomes available. For this reason, applications of Bayesian methods abound in areas such as adaptive filter theory [52] and neural networks [74].

The posterior density may also be used for parameter estimation. This is known as *Maximum a posteriori* (MAP) estimation. The supremum of the posterior density is located in parameter space using some suitable technique. From equation 2.9 one can see that a MAP estimate depends on the likelihood function weighted by the prior probability.

2.4 Prior probabilities

In this section, we discuss the selection of a prior probability density to describe one's state of knowledge (or lack of it!) about the value of a parameter before the data is observed.

2.4.1 Flat priors

One can claim to have no knowledge at all about the value of the parameter prior to observing the data. This state of ignorance may be described by using a prior probability density that is very broad and flat relative to the expected likelihood function.

The most intuitively obvious non-informative prior density is a uniform density. This prior is typically used for discrete distributions or for unbounded real valued parameters (i.e. *location* parameters [13]).

$$p(\theta \mid I_k) = k \tag{2.10}$$

where k is a constant. Parameter estimates obtained by MAP estimation using a uniform prior are identical to those obtained using maximum likelihood.

Purists can argue with some justification that the uniform prior probability is not valid because it may not be normalizable (e.g. if $\theta \in \Re$). This criticism is especially pertinent in the case of model selection, since the evidence depends on the constant k in equation 2.10, and if k is arbitrary, we can conclude that model evidence is arbitrary!

Prior probabilities are non-informative if they convey ignorance of the values of the parameters before observing the data compared to the state of knowledge afterwards. In this sense, the prior pdf need only be diffuse[1] in relation to the likelihood function.

Bretthorst [13] argues for the adoption of a normalized Gaussian prior:

$$p(\theta \mid I_k) = \frac{1}{\sqrt{2\,\pi\,\delta^2}} \exp\left[-\frac{\theta^2}{2\,\delta^2}\right] \tag{2.11}$$

---

[1]This means that the prior pdf will have larger variance than the likelihood function.

The parameter δ is known as a *hyperparameter*. To indicate ignorance of the value of a parameter one should set δ to a reasonably large value, although one is then back to the original problem of this value also being arbitrary. An acceptable solution to this problem[2] will be presented in section 2.6 where Bayesian model selection is discussed.

In the case of parameter estimation the normalizing factor does not actually matter since we are essentially interested in the *shape* of the posterior density and the location of the supremum. In this context, Bretthorst [13] shows that the parameter estimates obtained using a uniform prior probability are identically the same as those obtained using a Gaussian prior in the limit as the hyperparameter approaches infinity. In practice, as found by Pope [94], even a relatively small value of δ makes little difference to the estimate.

The situation with *scale parameters* is slightly different. A scale parameter, as its name suggests, is a measure of scale or magnitude. A scale parameter is therefore strictly positive. Jeffreys [58] discusses scale parameters and proposes that the prior pdf be uniformly distributed over different scales. This is the same as assuming that the logarithm of a scale parameter is uniformly distributed:

$$p\,(\log \sigma \mid I_k) = k \tag{2.12}$$

Using the fundamental transformation law of probabilities [89, 90] we can write

$$p\,(\log \sigma)\, d \log \sigma \propto p\,(\sigma)\, d\sigma \tag{2.13}$$

from which it follows that

$$p\,(\sigma \mid I_k) = \frac{k}{\sigma} \tag{2.14}$$

Not surprisingly, Jeffrey's prior is non-normalizable and is therefore improper. It can be made into a proper probability by placing bounds on the range:

$$p\,(\sigma \mid I_k) = \begin{cases} 0 & \text{if } \sigma < L_\sigma \\ 1/(\sigma \log (H_\sigma / L_\sigma)) & \text{if } L_\sigma \leq \sigma \leq H_\sigma \\ 0 & \text{if } \sigma > H_\sigma \end{cases} \tag{2.15}$$

Note that the normalizing factor in equation 2.15 often depends only very weakly on the bounds of the scale parameter.

2.4.2 Smoothness priors

The usual connotation associated with a non-informative prior is as a probability density of relatively large variance chosen to express ignorance of

---

[2] The hyperparameters are treated as nuisance parameters and integrated out.

the value of the parameters. Another basis for choosing a prior probability is to be conservative about the nature of the fit to the data that one expects to obtain after carrying out a Bayesian analysis.

Gersch and Kitigawa [37] and Rajan [99] have proposed smoothness priors for the analysis of time series. MacKay [74] has investigated the link between the theory of regularization, smoothness priors and Bayesian parameter estimation. MacKay's approach was to apply a regularizing prior probability, especially chosen to favour smooth model fits to the data. The likelihood, for its part, minimises the error of fit. The Bayesian MAP estimate combines the likelihood and the prior probability to achieve a compromise between accuracy and smoothness of fit. MacKay applies the theory to the interpolation of noisy data. This is a classic example of where accuracy in the fit to the data can lead to ridiculous estimates for the unknowns, which in this case are the missing samples (the interpolant).

Skilling [117] has derived an entropic prior pdf for use in image restoration. The prior probability favours entropy in the image[3], whereas the likelihood favours closeness of fit between the observed image and the restoration. Classical Bayesian image restoration using this prior was only moderately successful. Recently, theoretical justification has been established for using a Dirichlet prior [11] density and some present work on image restoration is based on using this prior density instead.

2.4.3 Convenience priors

Another criterion for the choice of a prior is its convenience.

In many situations the likelihood function belongs to the exponential family [126] of probability distributions. For ease of analysis, one would prefer the prior density to be *conjugate* to the likelihood function so the posterior density is also a member of the exponential family. Kheradmandnia [62] makes extensive use of the inverse gamma density as a prior density for scale parameters such as standard deviation. Conjugate priors are not necessarily non-informative.

The Jeffreys' priors and uniform priors discussed earlier are also examples of convenience priors. In the work presented later, non-informative priors such as Jeffreys' prior will be used.

2.5 The removal of nuisance parameters

One of the more interesting features of the Bayesian paradigm is the ability to remove nuisance parameters (i.e. parameters that are of no interest) from

---

[3]Entropy tends to favour smooth pdfs and in the case of image restoration encourages statistical independence between the pixels.

the analysis. Consider a particular problem involving two parameters a and b, where we are interested solely in inferring the value of a independently of the nuisance parameter b. Let us integrate the posterior density with respect to b as follows:

$$p(a \mid \mathbf{d}, I) = \int p(a, b \mid \mathbf{d}, I)\, db \tag{2.16}$$

The left hand side is a function of a alone and is often used to infer its value.

Marginalization (i.e. marginal inference) is an appealing procedure when the integral in equation 2.16 can be computed in closed form. In such cases one obtains a reduction in dimensionality since b is no longer present on the left hand side. The integrals for the general linear model discussed in section 2.7 fall into this class. As we will see, in those cases where the integrals must be computed numerically there is no real gain from the computational point of view of reducing dimensionality, because the integral has to be specified in terms of samples from the joint posterior density.

The marginal density gives the "whole picture" for a particular parameter in visual form. Using a marginal density is not equivalent [98] to using the joint density but it is certainly more informative than the standard practice of quoting an estimate and its associated error bars.

Note that marginalization necessitates a loss of information. Integrating through the posterior density in the manner of equation 2.16 means that all information about the value of b is lost. Some authors [13, 18, 74] advise estimating the value of a parameter from a marginal density (marginal point inference), when that marginal density is in closed form, and "back-substituting" this value into the joint density to estimate the second parameter (for which the marginal density is not in closed form). This procedure, although it works well for some problems (such in the estimation of the constituent frequencies of a data set), it is potentially disastrous in others.

The only case when marginal inference and joint inference are completely equivalent is when the joint posterior density is a completely separable function of the model parameters. In many examples, the joint posterior density is separable to a very good approximation, so the agreement between marginal point inference and joint point inference can be very good.

2.6 Model selection using Bayesian evidence

As noted earlier, the Bayesian evidence $p(\mathbf{d} \mid I_k)$ may be used to select signal models [19] and noise statistics [107] appropriate to the observed data. Evidence is defined by the following multiple integral:

$$p(\mathbf{d} \mid I_k) \quad = \quad \int_{\Theta} p(\mathbf{d} \mid \theta, I_k)\, p(\theta \mid I_k)\, \mathbf{d}\theta \tag{2.17}$$

where Θ is the model parameter space, and I_k is the model structure.

The term I_k represents the joint assumption of *both* the noise statistics and the signal model, which is called the data model. Note that this integral involves integrating the likelihood multiplied by the prior over *all* the parameters in the data model. In the case of discrete distributions the integration is replaced with a summation.

Consider a set of possible data models labelled I_1, I_2, $I_3 \dots I_M$ proposed to describe a given set of observations. We can use Bayes' theorem to find the posterior density of each model given the data:

$$p(I_k \mid \mathbf{d}) = \frac{p(\mathbf{d} \mid I_k) p(I_k)}{p(\mathbf{d})} \tag{2.18}$$

where

$$p(\mathbf{d}) = \sum_{j=1}^{M} p(\mathbf{d} \mid I_j) p(I_j) \tag{2.19}$$

If all models are equally likely *a priori* then we have

$$p(I_k) = \frac{1}{M} \tag{2.20}$$

Therefore the posterior probability of a model is given by the relative evidence:

$$p(I_k \mid \mathbf{d}) = \frac{p(\mathbf{d} \mid I_k)}{\sum_{j=1}^{M} p(\mathbf{d} \mid I_j)} \tag{2.21}$$

This constitutes the evidence framework for the selection of signal models. It is important to realise that in terms of real data, the correct data model may not be in the set chosen. All that we can do is to is to compare the candidate models that we have defined to determine which models are more plausible.

The computation of evidence is non-trivial in most cases because it involves integrating the product of the likelihood and the prior pdf over all the parameters in the model. Bretthorst [13] and MacKay [74] employ Gaussian approximations to the product of likelihood times prior, which may be integrated analytically. This approach, as well as numerous Monte Carlo models, will be described in the following two chapters.

2.6.1 Ockham's razor

The evidence framework has a very interesting property in that it accomplishes a compromise between accuracy of fit to the data and the parametric complexity of the model. The factor that penalizes complexity (known as the "Ockham factor") occurs because of the influence of the prior density on the model inference.

The mechanism by which the Ockham effect operates in Bayesian model selection is not difficult to understand. Before the data is gathered our prior information dictates that the probability is distributed over a region in space. After the data is gathered the likelihood concentrates this belief into a smaller region in parameter space. The prior must integrate to one, but only a small proportion of the prior in that region where the likelihood function is significant will have any effect on the value for the evidence. For example, if the prior is twice as broad as the likelihood function along each parameter axis, and the dimensionality of the parameter space is D, then the Ockham factor is given by $k \propto 1/2^D$. Hence, the Ockham effect is proportional to the ratio of the volume of parameter space occupied by the posterior density (posterior accessible volume) to the volume of parameter space occupied by the prior (prior accessible volume). One can see that the Ockham effect grows more pronounced if the prior density grows more diffuse. In the limit the Ockham effect grows infinite.

As mentioned earlier, there exist other approaches to penalizing model complexity. The information theoretic approaches such as AIC [1] and MDL [105, 113] are standard in the sense that they have been extensively applied. Experimental comparisons [33, 94] of the different criteria have shown that the Bayes' Ockham factor is superior and that, if anything, AIC tends to give too high a model order.

2.7 The general linear model

In this section, we discuss an extremely important signal model which may be used in a very large number of applications. Any data which may be described in terms of a linear combination of basis functions with an additive Gaussian noise component satisfies the general linear model. We will encounter many examples later so it is extremely profitable to study it in some detail here.

Suppose the observed data may be described by a model of the form:

$$d(i) = \sum_{k=1}^{M} b_k g_k(i) + e(i) \qquad \text{if} \quad 1 \leq i \leq N \tag{2.22}$$

where $g_k(i)$ is the value of a time dependent model function $g_k(t)$ evaluated at time t_i.

This can be written in the form of a matrix equation

$$\mathbf{d} = \mathbf{G}\,\mathbf{b} + \mathbf{e} \tag{2.23}$$

where $\mathbf{d}$ is an $N \times 1$ matrix of data points and $\mathbf{e}$ is an $N \times 1$ matrix of i.i.d. Gaussian noise samples. The matrix $\mathbf{G}$ is of size $N \times M$. Each column of $\mathbf{G}$ is a basis function evaluated at each point in the time series and each element of the $M \times 1$ matrix $\mathbf{b}$ is a linear coefficient (each corresponding to a particular column of the matrix $\mathbf{G}$).

The likelihood function is given by

$$p(\mathbf{d} \mid \{\omega\}, \sigma, \mathbf{b}, \mathbf{I}) = \left(2\,\pi\,\sigma^2\right)^{-\frac{N}{2}} \exp\left[-\frac{\mathbf{e}^{\mathbf{T}}\mathbf{e}}{2\,\sigma^2}\right] \tag{2.24}$$

where $\{\omega\}$ denotes the parameters of the matrix of basis functions $\mathbf{G}$.

Our aim is to derive a function[4] which may be used to infer the value of $\{\omega\}$ from the data without inferring the values for the nuisance parameters, namely the linear parameters $\mathbf{b}$ and the standard deviation σ.

Substituting equation 2.23 into equation 2.24 gives

$$p(\mathbf{d} \mid \{\omega\}, \mathbf{b}, \sigma, \mathbf{I}) = \left(2\,\pi\,\sigma^2\right)^{-\frac{N}{2}} \exp\left[-\frac{(\mathbf{d}-\mathbf{G}\mathbf{b})^{\mathbf{T}}(\mathbf{d}-\mathbf{G}\mathbf{b})}{2\,\sigma^2}\right] \tag{2.25}$$

We can see that from the point of view of the linear parameters $\mathbf{b}$ the likelihood is in the form of the exponential of a quadratic form.

We assume uniform priors over each of the elements of the matrix $\mathbf{b}$. In Appendix A it is shown how to integrate the posterior density to obtain:

$$\begin{aligned} p(\{\sigma, \omega\} \mid \mathbf{d}, \mathbf{I}) &= \int_{\Re^M} p(\{\mathbf{b}, \omega, \sigma\} \mid \mathbf{d}, \mathbf{I})\, d\mathbf{b} \\ &\propto \frac{\left(2\,\pi\,\sigma^2\right)^{-\left(\frac{N-M}{2}\right)}}{\sqrt{\det\left(\mathbf{G}^{\mathbf{T}}\mathbf{G}\right)}} \exp\left[-\left(\frac{\mathbf{d}^{\mathbf{T}}\mathbf{d}-\mathbf{f}^{\mathbf{T}}\mathbf{f}}{2\,\sigma^2}\right)\right] \end{aligned} \tag{2.26}$$

where $\Re^M$ is an M dimensional space of real numbers and $\mathbf{f}$ is given by equation 2.30 below.

The standard deviation may be integrated out as a gamma integral:

$$\int_0^\infty x^{\alpha-1} \exp(-Q\,x)\, dx = \frac{\Gamma(\alpha)}{Q^\alpha} \tag{2.27}$$

Assigning Jeffreys' prior to σ we can integrate out σ to give:

$$p(\{\omega\} \mid \mathbf{d}, \mathbf{I}) \propto \frac{\left[\mathbf{d}^{\mathbf{T}}\mathbf{d} - \mathbf{d}^{\mathbf{T}}\mathbf{G}\left(\mathbf{G}^{\mathbf{T}}\mathbf{G}\right)^{-1}\mathbf{G}^{\mathbf{T}}\mathbf{d}\right]^{\frac{-(N-M)}{2}}}{\sqrt{\det\left(\mathbf{G}^{\mathbf{T}}\mathbf{G}\right)}} \tag{2.28}$$

This is in the form of a student-t distribution.

---

[4]Bretthorst [13] uses the term *sufficient statistic* to describe this function.

Note that expression 2.28 is a function of $\{\omega\}$ only. This means that there is no need to know about the standard deviation nor the values of the linear parameters in order to infer the values of $\{\omega\}$. Here the integrals have been done analytically so the dimensionality of the parameter space was reduced for each parameter integrated out. This reduction of the dimensionality is a property of Bayesian marginal estimates and is a major advantage in many applications. An example of where this property is useful is in the estimation of the constituent frequencies of a signal [13, 14, 15, 16, 18, 56], where the values for the amplitudes and the noise variance are sometimes considered to be of little or no interest.

2.8 Interpretations of the general linear model

2.8.1 *Features*

The Bayesian statistics derived in the last section have some interesting features.

We can maximize the likelihood with respect to $\mathbf{b}$ to obtain an estimate for the amplitudes:

$$\hat{\mathbf{b}} = \left(\mathbf{G}^{\mathbf{T}}\mathbf{G}\right)^{-1}\mathbf{G}^{\mathbf{T}}\mathbf{d} \tag{2.29}$$

This is the familiar least squares (minimum variance) estimate that is obtained from linear algebra. The least squares fit to the data is given by:

$$\mathbf{f} = \mathbf{G}\,\hat{\mathbf{b}} \tag{2.30}$$

We can maximize the integrated likelihood in equation 2.26 to obtain an estimate of the variance that does not depend on the coefficients $\mathbf{b}$.

$$\begin{aligned} \hat{\sigma}^2 &= \frac{1}{N-M}\left[\mathbf{d}^{\mathbf{T}}\mathbf{d} - \mathbf{d}^{\mathbf{T}}\mathbf{G}\left(\mathbf{G}^{\mathbf{T}}\mathbf{G}\right)^{-1}\mathbf{G}^{\mathbf{T}}\mathbf{d}\right] \\ &= \frac{1}{N-M}\left[\mathbf{d}^{\mathbf{T}}\mathbf{d} - \mathbf{f}^{\mathbf{T}}\mathbf{f}\right] \end{aligned} \tag{2.31}$$

In other words, the estimated variance equals the difference between the data energy $\mathbf{d}^{\mathbf{T}}\mathbf{d}$ and the signal energy $\mathbf{f}^{\mathbf{T}}\mathbf{f}$ divided by the number of degrees of freedom $N - M$.

2.8.2 *Orthogonalization*

Bretthorst [13] transforms the linear coefficients $\mathbf{b}$ to a different basis $\mathbf{a}$ such that

$$\mathbf{a}^{\mathbf{T}}\mathbf{a} = \mathbf{b}^{\mathbf{T}}\mathbf{G}^{\mathbf{T}}\mathbf{G}\,\mathbf{b} \tag{2.32}$$

This transformation converts $\mathbf{G}^{\mathbf{T}}\mathbf{G}$ into the identity matrix. This procedure is termed *orthogonalization* and the components of the parameter vector $\mathbf{a}$ are termed *orthogonalized amplitudes*. Bretthorst chooses to work with orthogonalized amplitudes $\mathbf{a}$ instead of the original coefficients $\mathbf{b}$ with which the model function was originally specified. The appeal in using this transformation is that it simplifies the posterior probability in expression 2.28, although it could be argued [97] that by using it a different problem is being solved.

2.9 Example of marginalization

As an example of the application of the general linear model, consider the detection of a single frequency.

$$f(t) = A\cos(\omega t) + B\sin(\omega t) \tag{2.33}$$

The data is assumed to consist of samples from the signal $f(t)$ corrupted with independent white zero mean Gaussian noise samples with standard deviation σ.

The signal model just described agrees with the general linear model. The structure of the $\mathbf{G}$ matrix is:

$$\mathbf{G} = \begin{bmatrix} \cos(\omega t_1) & \sin(\omega t_1) \\ \cos(\omega t_2) & \sin(\omega t_2) \\ \cos(\omega t_3) & \sin(\omega t_3) \\ \vdots & \vdots \\ \cos(\omega t_N) & \sin(\omega t_N) \end{bmatrix} \tag{2.34}$$

and the linear coefficient vector is:

$$\mathbf{b} = \begin{bmatrix} A \\ B \end{bmatrix} \tag{2.35}$$

The linear coefficients and the noise standard deviation may be integrated out as described previously to give the probability density for the angular frequency:

$$p(\omega \mid \mathbf{d}, \mathbf{I}) \propto \frac{\left[\mathbf{d}^{\mathbf{T}}\mathbf{d} - \mathbf{d}^{\mathbf{T}}\mathbf{G}\left(\mathbf{G}^{\mathbf{T}}\mathbf{G}\right)^{-1}\mathbf{G}^{\mathbf{T}}\mathbf{d}\right]^{\frac{2-N}{2}}}{\sqrt{\det\left(\mathbf{G}^{\mathbf{T}}\mathbf{G}\right)}} \tag{2.36}$$

Bretthorst [13] notes that the columns of the $\mathbf{G}$ matrix are nearly orthogonal. This may be exploited to simplify the marginal posterior for ω. The matrix $\mathbf{G}^{\mathbf{T}}\mathbf{G}$ is approximately equal to $N/2$ times the identity matrix. It is easy to show that:

$$\mathbf{d}^{\mathbf{T}}\mathbf{d} - \mathbf{f}^{\mathbf{T}}\mathbf{f} \approx \sum_{i=1}^{N} d_i^2 - \frac{2}{N}\left(\sum_{i=1}^{N} d_i\cos(\omega t_i) + \sum_{i=1}^{N} d_i\sin(\omega t_i)\right)^2 \tag{2.37}$$

Define

$$I(\omega) = \left(\sum_{i=1}^{N} d_i \cos(\omega t_i)\right) \tag{2.38}$$

$$Q(\omega) = \left(\sum_{i=1}^{N} d_i \cos(\omega t_i)\right) \tag{2.39}$$

We can make further approximations to equation 2.37 to obtain:

$$\mathbf{d}^{\mathbf{T}}\mathbf{d} - \mathbf{f}^{\mathbf{T}}\mathbf{f} \approx \sum_{i=1}^{N} d_i^2 - \frac{2}{N} I(\omega)^2 - \frac{2}{N} Q(\omega)^2 \tag{2.40}$$

Bretthorst [13] defines the Schuster periodogram as:

$$C(\omega) = \frac{1}{N}\left(I^2(\omega) + Q^2(\omega)\right) \tag{2.41}$$

We can write the marginal density for the angular frequency in terms of the Schuster periodogram as:

$$p(\omega \mid \mathbf{d}, \mathbf{I}) \propto \left[1 - \frac{2\,C(\omega)}{\sum_{i=1}^{N} d_i^2}\right]^{\frac{2-N}{2}} \tag{2.42}$$

Note that the term inside the square brackets is small (thus implying that the marginal density is large) if $2\,C(\omega) \approx \sum_{i=1}^{N} d_i^2$. This occurs if most of the data energy is concentrated around a single frequency ω.

2.9.1 Results

Consider the synthetic data plotted in figure 2.1. The data consists of 100 samples of a unit amplitude sine wave with an angular frequency of 10 rad s^{-1}. The signal is corrupted with additive zero mean i.i.d. Gaussian noise of unit variance.

The corresponding marginal density (not normalised) is shown in figure 2.2. It is composed of expression 2.42 evaluated for one hundred values of ω over the range of angular frequencies. Note that the point of maximum probabilty 9.95 rad s^{-1} (i.e. the point estimate) is not equal to the true value. The difference between the point estimate and the true value is a function of the realisation of the noise corrupting the data. However, the true value 10 rad s^{-1} lies well within the body of the marginal density, and it is therefore plausible that 10 rad s^{-1} is the single frequency that we seek.

There is a very interesting comparison to be made between the discrete Fourier transform power spectrum and the Bayesian marginal density for a single frequency. Figure 2.3 shows the logarithm of the marginal density

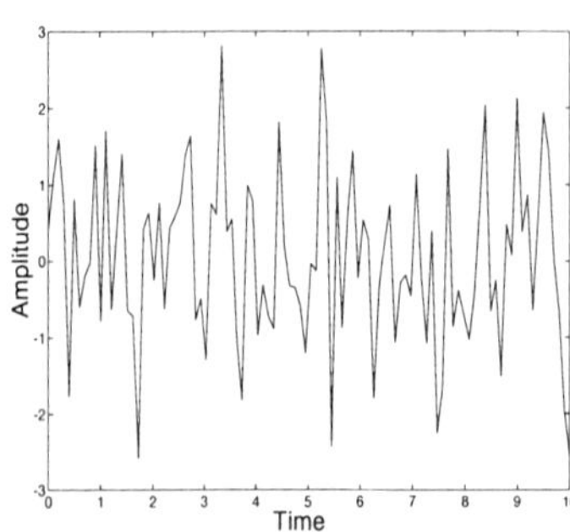

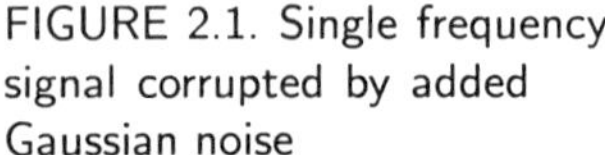

FIGURE 2.1. Single frequency signal corrupted by added Gaussian noise

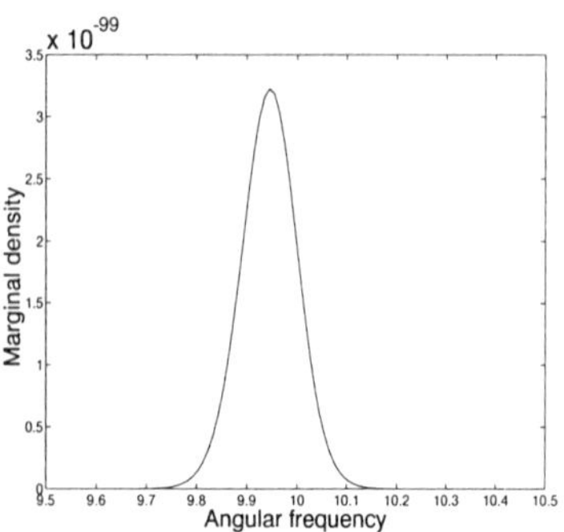

FIGURE 2.2. Marginal density for single frequency (not normalised)

for a single frequency. Figure 2.4 is the discrete FFT power spectrum, computed using 2048 bins. The similarity between the shape of the two graphs is immediately noticeable. Bretthorst [14] shows that this is not a coincidence. If the frequency bin positions correspond with the sample points in the time series then the two procedures will be exactly equivalent. Bretthorst goes on to discuss the limitations of the FFT power spectrum as a frequency estimator, based on its Bayesian derivation. The marginal density in 2.36, and hence the Schuster periodogram, is designed to determine the value of a single frequency in white Gaussian zero mean noise. Therefore it should really only be used on data that satisfies the single frequency model, and is not really designed for use in resolving two or more closely spaced frequencies, nor should it be used when the data is corrupted by non-Gaussian noise.

2.10 Example of model selection

In this section we shall present two simple examples in which Bayesian evidence can be computed in closed form. The aim in each case is to determine the appropriate model order for the data set under consideration. The two examples will be to determine the most probable order of a polynomial fit to noisy data and to determine the order of an autoregressive process. Both polynomials and autoregressive models satisfy the general linear model, differing only in the choice of the matrix $\mathbf{G}$ in equation 2.28.

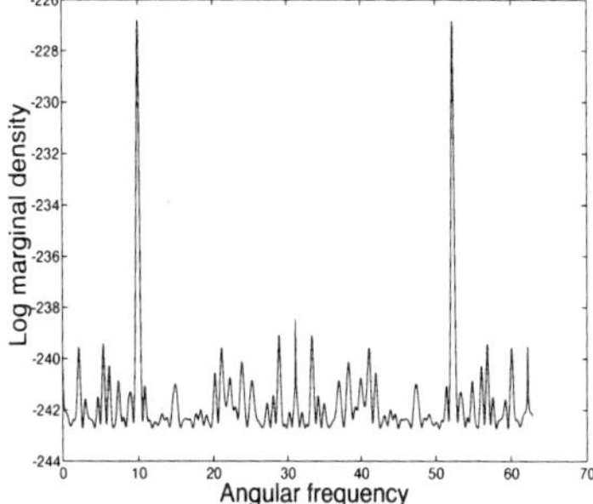

FIGURE 2.3. Log marginal density for single frequency

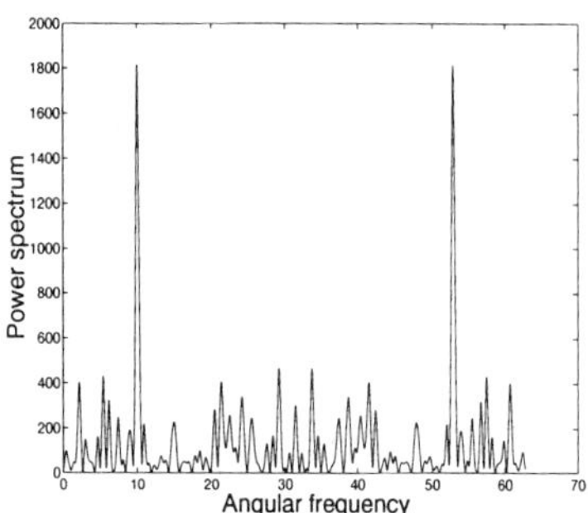

FIGURE 2.4. Discrete FFT power spectrum (2048 bins)

2.10.1 Closed form expression for evidence

The evidence for a general linear model, in which the **G** matrix has no parameters, will now be derived.

The first stage of the analysis is to choose normalized prior densities for the parameters. Bretthorst [13] argues for the adoption of a Gaussian prior for **b**.

$$p(\mathbf{b} \mid \delta, I) = \left(2\pi\delta^2\right)^{-M/2} \exp\left[-\frac{\mathbf{b}^{\mathbf{T}}\mathbf{b}}{2\,\delta^2}\right] \tag{2.43}$$

The parameter δ is a hyperparameter.

We shall assume the normalized Jeffreys' prior in expression 2.14 for the hyperparameter δ and the standard deviation σ. Let R_δ and R_σ be the respective normalization constants.

The evidence is given by the following integral:

$$p(\mathbf{d} \mid I) = \int\int\int p(\mathbf{d} \mid \mathbf{b}, \sigma, I)\, p(\sigma \mid I)\, p(\mathbf{b} \mid \delta, I)\, p(\delta \mid I)\, d\mathbf{b}\, d\delta\, d\sigma \tag{2.44}$$

where the likelihood is defined in equation 2.25, and the limits of integration are $\{L_\sigma < \sigma < H_\sigma, L_\delta < \delta < L_\delta \text{ and } \mathbf{b} \in \Re^M\}$.

This integral is not straightforward to compute as it stands, so we will introduce some simplifying assumptions. First, we assume that the supremum of the likelihood function lies within the range of the prior probabilities on δ and σ. Second, we also assume that the bounds on the prior probability for σ and δ lie out in the tails of the likelihood function. The contribution to the integral caused by the cutoff should be negligible and may therefore be ignored. Hence we replace the bounds in equation 2.44 with

$$H_\sigma = \infty \quad L_\sigma = 0$$
$$H_\delta = \infty \quad L_\delta = 0$$

without changing the prescribed values for R_δ and R_σ. This allows us to integrate out δ and σ using the gamma integral in equation 2.27.

We also assume that the prior probability density for $\mathbf{b}$ is diffuse relative to the likelihood. The probability mass is mainly concentrated about the supremum of the posterior density, which means that to a good approximation one may treat the likelihood as a delta function when integrating out $\mathbf{b}$. The value of the amplitudes $\hat{\mathbf{b}}$ and the corresponding fit to the data $\mathbf{f}$ corresponding to the supremum of the posterior density are given by equation 2.29 and 2.30 respectively. Using these assumptions, the evidence may be shown to be:

$$p(\mathbf{d} \mid I) \approx \frac{\pi^{-N/2}\,\Gamma\left(\frac{M}{2}\right)\Gamma\left(\frac{N-M}{2}\right)\left|\mathbf{G}^{\mathbf{T}}\mathbf{G}\right|^{-1/2}}{4\,R_\delta\,R_\sigma\left(\hat{\mathbf{b}}^{\mathbf{T}}\hat{\mathbf{b}}\right)^{M/2}\left(\mathbf{d}^{\mathbf{T}}\mathbf{d}-\mathbf{f}^{\mathbf{T}}\mathbf{f}\right)^{(N-M)/2}} \tag{2.45}$$

This approximation to the evidence may be applied to any model that satisfies expression 2.22.

2.10.2 Determining the order of a polynomial

Consider the synthetic time series plotted in figure 2.5. This time series is generated by evaluating a polynomial $p(x)$ at one hundred uniformly spaced abscissae over the range $[0,1]$ and adding white zero mean Gaussian noise with standard deviation $\sigma = 0.001$.

$$p(x_i) = (x_i - 2)(x_i - 1)(x_i - 0.5)\,x_i \tag{2.46}$$

Figure 2.6 is a plot of the log evidence of different orders of polynomial fit to the data in figure 2.5. The log evidence increases to a maximum at order $p = 4$. This maximum is the compromise between the conflicting objectives of fitting the data, as dictated by the likelihood function, which in the case of polynomials always improves with increasing model order, and the need for parametric simplicity in the model, as dictated by the prior.

Figure 2.7 is a plot of the relative probability of each model order. In this case the model order with highest probability is $p = 4$ which corresponds to the "truth": the model order used when generating the data.

2.10.3 Determining the order of an AR process

Consider the synthetic time series plotted in figure 2.8. This time series is a realization of an eight pole autoregressive process and consists of 300 samples. The eight poles are positioned as conjugate pairs at $-0.4572 \pm j\,0.7765$, $-0.8284 \pm j\,0.1079$, $-0.5979 \pm j\,0.6138$ and $-0.5976 \pm j\,0.5364$.

Figure 2.9 shows the log evidence for each order of AR model to a maximum of 20. Note that the log evidence monotonically increases until it

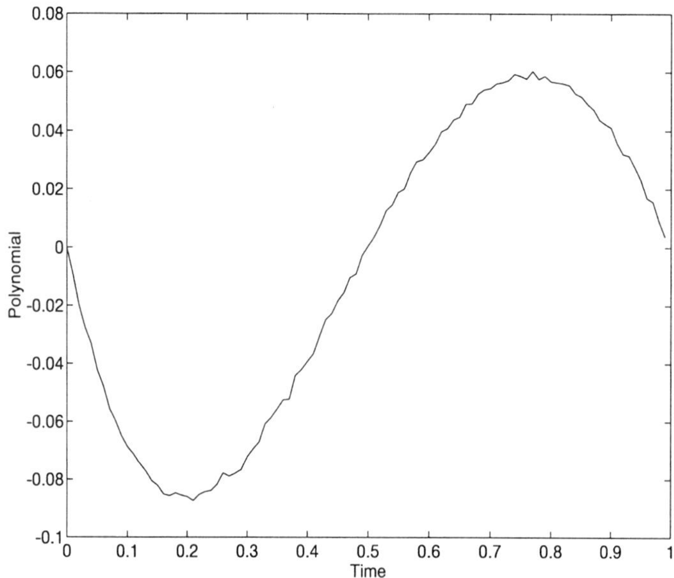

FIGURE 2.5. Fourth order polynomial data

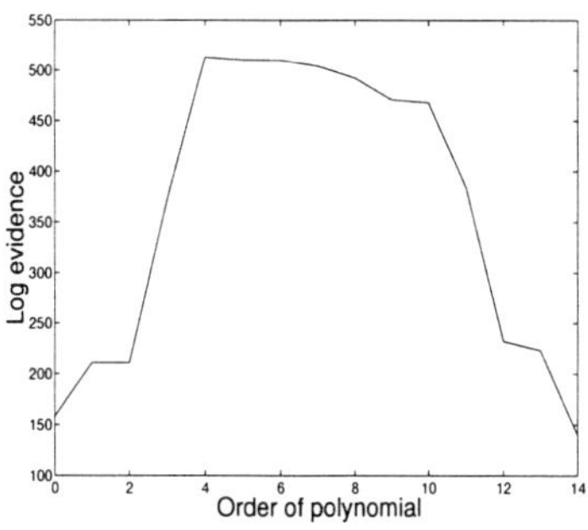

FIGURE 2.6. Log evidence for polynomial data

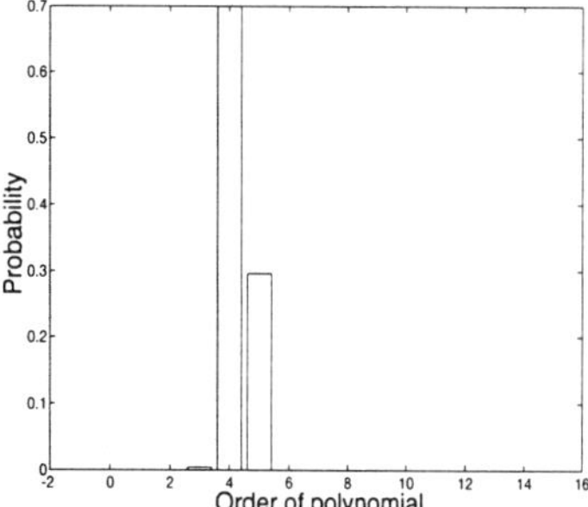

FIGURE 2.7. Probable order for polynomial data

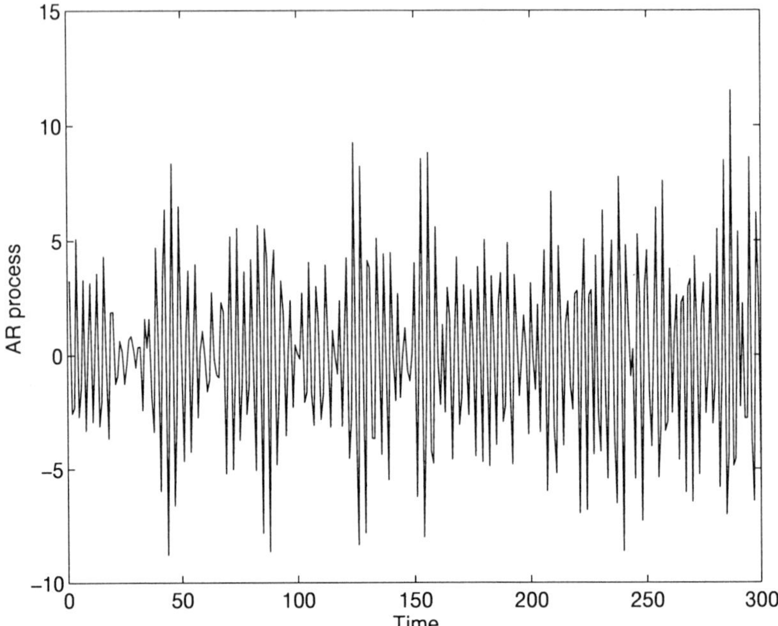

FIGURE 2.8. Eight pole autoregressive data

reaches a plateau when the model order reaches $p = 8$. If one assumes that the true model order lies in the range $1 < p < 20$ then the probability of each model can be evaluated in turn as shown in figure 2.10.

The result of Bayesian inference is that there is a 85% probability that the model order of the AR data in figure 2.8 is 8.

2.11 Concluding remarks

This chapter has introduced the Bayesian formalism. The likelihood function was defined and prior probabilities were introduced. Using Bayes' theorem, the likelihood function was combined with the prior to give the posterior density. It was then shown how to carry out parameter estimation using marginal densities and model selection using Bayesian evidence. The model selection procedure was shown to exhibit the Ockham effect, by which parametrically simple models are favoured over more complex ones.

The general linear model was developed in a compact matrix form. This signal model assumes that the signal consists of a linear combination of time

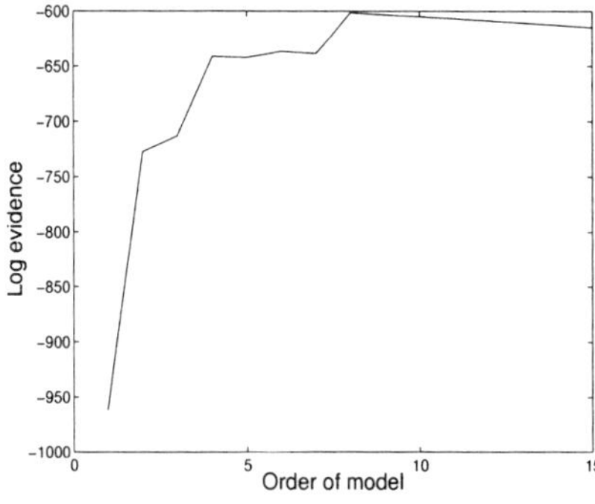

FIGURE 2.9. Log evidence for autoregressive data

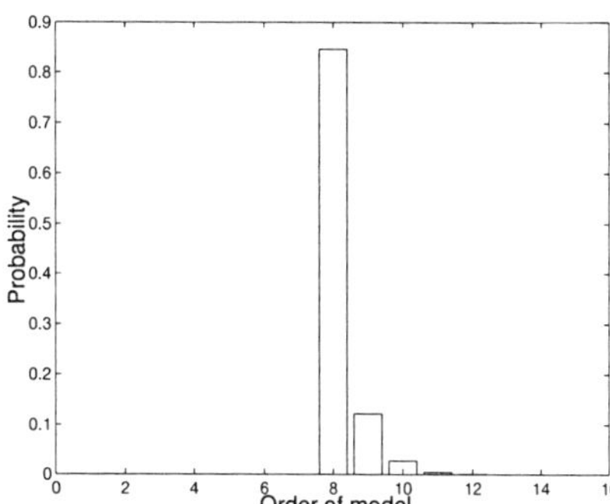

FIGURE 2.10. Probable order for autoregressive data

dependent basis functions, and is widely applicable in signal processing and data analysis. It was also demonstrated how nuisance parameters, such as the standard deviation of the Gaussian noise and the coefficients of the basis functions, can be removed by integrating the posterior density.

By assuming priors such as uniform or Jeffreys' priors, or even just that the prior is an extremely smooth function in relation to the likelihood, it was possible to compute integrals for evidence and marginal densities in closed form. At this stage it is obvious that the Gaussian general linear model is a special case and that, in general, Bayesian integrals may not normally be as tractable as the examples presented in this chapter.

Most non-Gaussian noise environments or models involving non-linear parameters involve posterior densities that are not integrable in closed form. In future chapters, numerical methods will be introduced to carry out Bayesian inference in such situations.

3

Numerical Bayesian Inference

The aim of this chapter is to provide an extensive survey of numerical Bayesian methods. The emphasis is placed primarily on the numerical integration of posterior densities to obtain approximations to marginal densities and Bayesian evidence.

It will be assumed that the mass of the probability density is concentrated about a single dominant mode. This is a reasonable assumption to make about most problems of interest. It is almost in the nature of probabilistic inference that one is concerned with those situations where the optimum solution[1] is concentrated over a single region in parameter space and that one is interested in the location and extent of this region.

The structure of this chapter is as follows. The chapter begins with a discussion of the behaviour of the log likelihood with an increasing number of data. This leads to the principle of asymptotic normality. This principle forms the basis for many methods that may be used for making numerical Bayesian analyses more tractable. These methods include the use of suitable reparameterization and the orthogonalization of parameter spaces. The principle is extended to compute non-normal approximations to marginal densities based on saddlepoint methods and the Laplace approximation.

A brief survey of both local and global optimization techniques is presented. The optimization of probability densities may be carried out for its own sake for MAP parameter estimation, but also for its usefulness in numerical integration. Both optimization and numerical integration may

---

[1]More precisely the *highest posterior density region.*

be carried out using methods based on random sampling. As we proceed, the distinctions between the apparently disparate areas of numerical integration, optimization and random simulation will become increasingly blurred.

The theory behind the computation of marginal densities and evidence using Monte Carlo methods is developed. This includes a novel approach for computing marginal densities known as the dummy variable method, a special case of which corresponds to a very important standard technique. A technique for evaluating the evidence of a model is also devised.

3.1 The normal approximation

The posterior pdf was defined in equation 2.9 to be likelihood multiplied by prior pdf divided by a normalization constant known as the evidence. An understanding of the nature of the posterior pdf is important since knowledge about this function may suggest effective ways of computing integrals such as marginal densities and evidence.

We will soon see that under some circumstances the posterior density may be well approximated by a Gaussian form. The first case that we will study is the phenomenon known as asymptotic normality. It can be shown [138] that if the posterior density satisfies certain regularity conditions then the posterior density will asymptotically approach normality as the number of data increases.

The principle of asymptotic normality can be very useful in Bayesian applications. Although it is only provides a guide to the asymptotic behaviour of the posterior density, it can be reasonably accurate even for moderate sample size [131]. For small sample size, it will usually not be at all accurate but it can at least provide a useful starting point for a full Bayesian analysis. As noted by Smith [119], Gaussianity can also be enforced by reparameterization.

To summarize, the normal approximation is interesting for two reasons. First, it provides an approximation to the posterior density that may be integrated quite readily to approximate marginal densities and evidence. Second, it can give a starting point from which difficult non-normal densities can be analysed.

3.1.1 Effect of number of data on the likelihood function

The behaviour of the posterior density is usually strongly determined by the likelihood. The likelihood function itself is strongly affected by the number of data.

Consider the log pdf for Gaussian data with error sequence $\mathbf{e}$:

$$\log p(\mathbf{e}) = -\frac{\mathbf{e}^{\mathbf{T}}\mathbf{e}}{2\,\sigma^2} - \frac{N}{2}\log\left(2\,\pi\,\sigma^2\right) \tag{3.1}$$

where N is the sample size. The variance may be estimated from the data as

$$\sigma^2 \approx -\frac{\mathbf{e}^{\mathbf{T}}\,\mathbf{e}}{N} \tag{3.2}$$

Substituting σ^2 into equation 3.1 we obtain the log MAP probability

$$\log\,\hat{p} \approx -\frac{N}{2} - \frac{N}{2}\log\left(2\,\pi\,\sigma^2\right) \tag{3.3}$$

This means that the log probability density is approximately linear in the number of data. The slope depends on the standard deviation. The generalization of this result to densities other than Gaussian is straightforward.

3.1.2 Taylor approximation

Consider the following Taylor approximation to the log density:

$$l\,(\theta) \approx l\,(\hat{\theta}) + (\theta - \hat{\theta})^{\mathbf{T}}\,\nabla^{\mathbf{T}} l(\theta)\Big|_{\theta=\hat{\theta}} + \frac{1}{2}(\theta - \hat{\theta})^{\mathbf{T}}\,\nabla\nabla^{\mathbf{T}}\,l(\theta)\,\Big|_{\theta=\hat{\theta}}(\theta - \hat{\theta}) \tag{3.4}$$

where $N\,l\,(\theta) = \log p\,(\theta)$.

If $\hat{\theta}$ is a mode of the joint density then we have $\nabla l(\hat{\theta}) = 0$. Taking exponentials of equation 3.4 gives

$$p\,(\theta) \approx p\,(\hat{\theta})\exp\left[\frac{1}{2}(\theta - \hat{\theta})^{\mathbf{T}} N\;\nabla\nabla^{\mathbf{T}}\,l(\theta)\,\Big|_{\theta=\hat{\theta}}(\theta - \hat{\theta})\right] \tag{3.5}$$

This is in the form of a Gaussian density with the inverse covariance matrix equal to minus the Hessian matrix:

$$C^{-1} = -N\;\nabla\nabla^{\mathbf{T}}\,l(\theta)\,\Big|_{\theta=\hat{\theta}} \tag{3.6}$$

The Taylor expansion in equation 3.4 is only accurate within a small region in parameter space about the mode. As the log likelihood is scaled by increasing numbers of data then the likelihood itself becomes increasingly peaked with the curvature of the log likelihood at the mode increasing of the order $\mathcal{O}(N^{1/2})$. As a result the probability mass becomes concentrated about the global maximum of the posterior density. This is the region where the Taylor expansion gives a good approximation, and thus the posterior density is asymptotically normal with accuracy in the order $\mathcal{O}(N^{1/2})$. A full mathematical treatment is presented by Walker [138].

3.1.3 Reparameterization

Even in those cases where the density is highly non-normal, it is possible to make it more normal by means of some suitable parameterization. The first requirement of any parameterization is that the parameter extends over the real axis. The second requirement is that in the case of thick tailed densities that one should narrow the tails in order that the reparameterized density resembles a Gaussian.

Gamma density

As an example consider the gamma form

$$g(x) = x^{t-1} \exp(-x) \tag{3.7}$$

where x is strictly positive. Suppose we are interested in finding the normalization constant for $g(x)$ which we will denote as G:

$$G = \int_0^\infty g(x)\, dx \tag{3.8}$$

The normalization constant is the gamma function of t. It is our aim to find an approximation to this function.

The Gaussian approximation to expression 3.7 is:

$$g(x) \approx (t-1)^{(t-1)} \exp\left[-\frac{(x-(t-1))^2}{2(t-1)}\right] \exp[-(t-1)] \tag{3.9}$$

Integrating over the *whole* real axis gives:

$$G_1 = \sqrt{2\pi}\,(t-1)^{t-1/2} \exp[-(t-1)] \tag{3.10}$$

A better approximation may be obtained by letting $y = \log x$. The integral then becomes:

$$G = \int_{\Re} \exp[ty] \exp[-e^y]\, dy = \int_{\Re} g(x)\, x\, d\log x \tag{3.11}$$

Making a normal approximation to the integrand it is easy to show that:

$$G \approx \exp[t \log t - t] \int_{\Re} \exp\left[-\frac{(y-\log t)^2}{2t}\right] dy \tag{3.12}$$

Integrating the normal approximation gives:

$$G_2 \approx \sqrt{2\pi}\, t^{t-1/2} \exp(-t) \tag{3.13}$$

This is the well known *Stirling approximation* [27] to the gamma function.

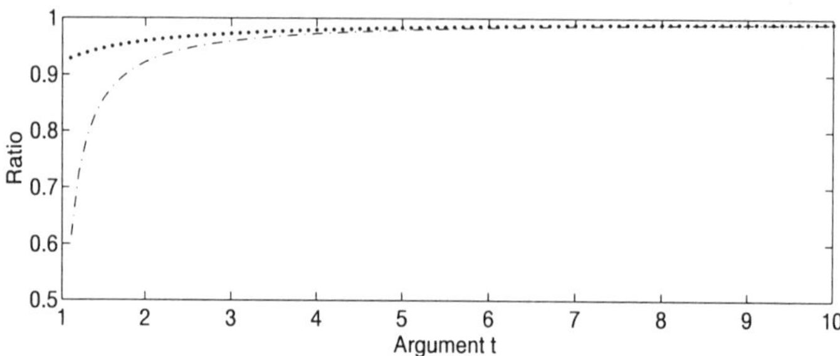

FIGURE 3.1. Approximations to the gamma function: (Dotted) Stirling's approximation; (Dashed) Unparameterized Gaussian approximation

The performance of the two approximations is assessed in figure 3.1. The X-axis is the argument t and the Y-axis is the ratio of the approximation to the gamma function. The approximation 3.13 shown as a dotted curve is superior to 3.10 which is shown as a dashed curve. As expected, the beneficial effects of reparameterization are more noticeable for smaller values of t.

General reparameterization

Smith [119] recommends a log transformation for positive scale parameters such as x in equation 3.7, and in particular for the variance parameters of a probability density. Smith also recommends a logit transformation[2] for variables bounded between zero and one, which is very useful for proportions.

In the case of multivariate densities it is also extremely effective to carry out a further linear transformation on the transformed parameters to form an orthogonal set. Figure 3.2 is a bivariate Gaussian density function. Figure 3.3 shows the same bivariate Gaussian density after an orthogonalizing transformation has been applied.

The combination of simple functional transformations and orthogonalization can be extremely effective at rendering difficult densities tractable. A particularly spectacular example of the beneficial effects of parameterization (the probit example) is given by Smith *et al.* [120].

Hills and Smith [49, 50] have developed a method known as the Bayes t-plot for the suitable parameterization of awkward densities.

---

[2] $\text{logit}(x) = \log\left[x/(1-x)\right]$

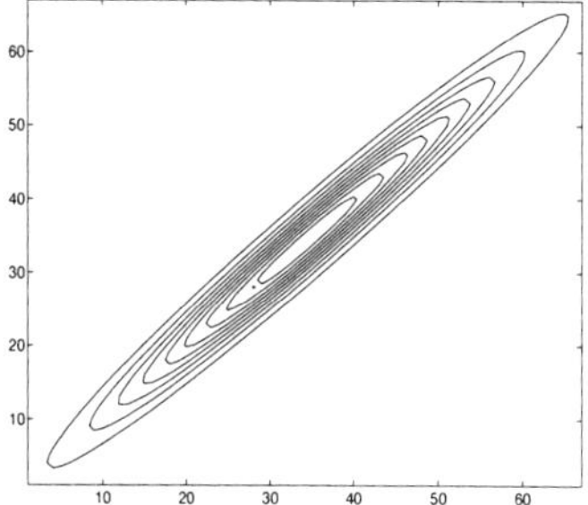

FIGURE 3.2. Untransformed density

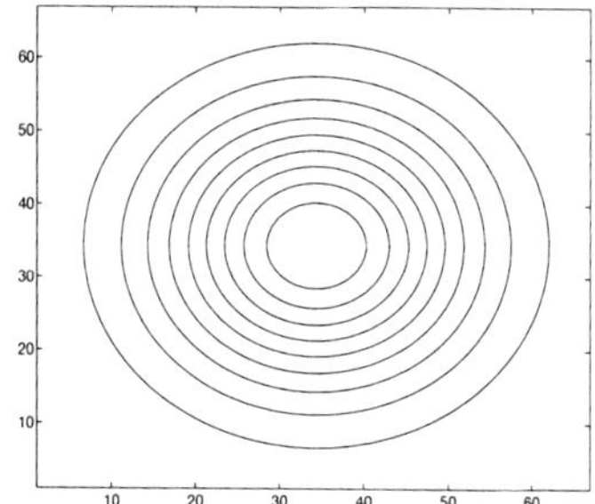

FIGURE 3.3. Orthogonalized density

3.1.4 Jacobian of transformation

Let $\mathbf{x}$ be the parameter vector and $\mathbf{y} = \mathbf{f}(\mathbf{x})$ be a prescribed function, possibly designed to make the product of likelihood times prior more Gaussian. The integral may be written as

$$\int_{\mathbf{X}} p(\mathbf{x})\, d\mathbf{x} = \int_{\mathbf{Y}} p\left(f^{-1}(\mathbf{y})\right) \left|\frac{\partial \mathbf{x}}{\partial \mathbf{y}}\right| d\mathbf{y} \tag{3.14}$$

where $\left|\frac{\partial \mathbf{x}}{\partial \mathbf{y}}\right|$ is the Jacobian of the transformation.

Note that reparameterization introduces an additional Jacobian term into the integrand. It is no longer the case that one is integrating the product of likelihood times prior but rather likelihood times prior times Jacobian. To avoid confusion in future, we shall always refer to this product as the integrand.

In general, the likelihood dominates the behaviour of the integral because it is far more narrowly peaked than the prior probability (assuming a non-informative prior). For all intents and purposes the likelihood will usually resemble a delta function [13] when compared to the prior.

3.1.5 Normal approximation to evidence

We have already seen that in certain circumstances, such as when the sample size is very large or under the effects of a suitable parameterization, that the joint density will be well approximated by a normal density. This leads to a simple and rather elegant approximation to the evidence.

Let us assume that the integrand is well approximated by a Gaussian

$$p(\theta) \approx p(\hat{\theta}) \exp\left[\frac{1}{2}(\theta - \hat{\theta})^{\mathbf{T}} \nabla\nabla^{\mathbf{T}}\, l(\theta)\Big|_{\theta=\hat{\theta}} (\theta - \hat{\theta})\right] \tag{3.15}$$

where $l(\theta) = \log p(\theta)$.

The evidence is the integral defined by expression 2.17. Note that the integral is performed with respect to *all* the parameters

$$\begin{aligned} E &= \int_{\Re^p} p(\theta)\, d\theta \\ &\approx \int_{\Re^p} p(\hat{\theta}) \exp\left[\frac{1}{2}(\theta-\hat{\theta})^{\mathbf{T}} \nabla\nabla^{\mathbf{T}}\, l(\theta)\Big|_{\theta=\hat{\theta}} (\theta-\hat{\theta})\right] d\theta \\ &= p(\hat{\theta}) \frac{(2\,\pi)^{p/2}}{\sqrt{\left|\det\, \nabla\nabla^{\mathbf{T}}\, l(\theta)\Big|_{\theta=\hat{\theta}}\right|}} \end{aligned} \quad (3.16)$$

This approximation is extensively used by MacKay [74] among others. As expected, it is a good approximation when the number of data is large.

3.1.6 Normal approximation to the marginal density

In this section we discuss the normal approximation to the marginal density. We begin by partitioning parameter space $\mathbf{Z}$ into two subspaces $\mathbf{X}$ and $\mathbf{Y}$.

$$f(\mathbf{x}) = \int_{\mathbf{Y}} f(\mathbf{x},\, \mathbf{y})\, \mathbf{dy} \quad (3.17)$$

Let us make a normal approximation to the posterior density $f(\mathbf{x},\,\mathbf{y})$ as follows:

$$f(\mathbf{x},\,\mathbf{y}) \approx f(\hat{\mathbf{x}},\,\hat{\mathbf{y}}) \exp\left[-\frac{1}{2}(\mathbf{z}-\hat{\mathbf{z}})^{\mathbf{T}}\, \mathbf{C}^{-1}\, (\mathbf{z}-\hat{\mathbf{z}})\right] \quad (3.18)$$

where $\mathbf{z} \in \mathbf{Z}$ is a vector in parameter space with components $\mathbf{x} \in \mathbf{X}$ and $\mathbf{y} \in \mathbf{Y}$. The inverse covariance matrix may be estimated by numerical differentiation of the log density at the mode.

Let us partition the inverse covariance matrix as follows:

$$\mathbf{C}^{-1} = \begin{bmatrix} \mathbf{A} & \mathbf{B} \\ \mathbf{B}^{\mathbf{T}} & \mathbf{D} \end{bmatrix} \quad (3.19)$$

so the exponent in equation 3.18 may be written as follows:

$$\begin{aligned} (\mathbf{z}-\hat{\mathbf{z}})^{\mathbf{T}}\, \mathbf{C}^{-1}\, (\mathbf{z}-\hat{\mathbf{z}}) &= (\mathbf{x}-\hat{\mathbf{x}})^{\mathbf{T}}\, \mathbf{A}\, (\mathbf{x}-\hat{\mathbf{x}}) + 2\, (\mathbf{x}-\hat{\mathbf{x}})^{\mathbf{T}}\, \mathbf{B}\, (\mathbf{y}-\hat{\mathbf{y}}) \\ &\quad + (\mathbf{y}-\hat{\mathbf{y}})^{\mathbf{T}}\, \mathbf{D}\, (\mathbf{y}-\hat{\mathbf{y}}) \end{aligned} \quad (3.20)$$

If we take the subspace $\mathbf{X} = \Re^{p-m}$ and $\mathbf{Y} = \Re^m$ then we can substitute equation 3.20 into equation 3.18 to obtain:

$$\begin{aligned} f(\mathbf{x}) &= \int_{\mathbf{Y}} f(\mathbf{x},\, \mathbf{y})\, dy \\ &\approx \frac{2\,\pi^{(p-m)/2}}{\sqrt{\det \mathbf{D}}} \exp\left[-\frac{1}{2}(\mathbf{x}-\hat{\mathbf{x}})^{\mathbf{T}} \left(\mathbf{A}-\mathbf{B}^{\mathbf{T}}\, \mathbf{D}^{-1}\, \mathbf{B}\right) (\mathbf{x}-\hat{\mathbf{x}})\right] \end{aligned} \quad (3.21)$$

In other words, the marginal posterior density is Gaussian with inverse covariance matrix given by

$$\mathbf{C}_{\mathbf{X}}^{-1} = \mathbf{A} - \mathbf{B}^{\mathbf{T}}\,\mathbf{D}^{-1}\,\mathbf{B} \tag{3.22}$$

In fact the inverse covariance matrix may be derived using a less circuitous route than above. Press *et al* [95] consider the problem of obtaining the inverse of part of a matrix from the full inverse. Let us write the covariance matrix $\mathbf{C}$ as follows:

$$\mathbf{C} = \begin{bmatrix} \tilde{\mathbf{A}} & \tilde{\mathbf{B}} \\ \tilde{\mathbf{B}} & \tilde{\mathbf{D}} \end{bmatrix} \tag{3.23}$$

Using the formulae given by Press *et al* [95] it is easy to show that:

$$\tilde{\mathbf{A}}^{-1} = \mathbf{A} - \mathbf{B}^{\mathbf{T}}\,\mathbf{D}^{-1}\,\mathbf{B} \tag{3.24}$$

Therefore one can write:

$$\mathbf{C}_{\mathbf{X}} = \tilde{\mathbf{A}} \tag{3.25}$$

The covariance matrix of the approximation to the marginal posterior density is a partition of the entire covariance matrix.

There are some interesting implications to equation 3.24. The left hand side is the curvature of the Gaussian approximation to the marginal density. The right hand side is the curvature of the Gaussian approximation to the joint density minus some positive semidefinite term. In other words, the approximation to the marginal density is always as curved or *less* than the approximation to the joint density. Therefore in the limit as the number of data approach infinity, one can deduce that marginal densities make more conservative estimates than joint densities.

3.1.7 The delta method

It often occurs that one is more interested in finding the marginal density of a function of a parameter rather than of the parameter itself. One example of where this may occur is when we have transformed the parameters to a more "orthogonal" basis for the sake of ease of integration, but where we wish to obtain the posterior density of an untransformed parameter. Stated more formally, given the joint posterior $p(\theta)$ we desire a Gaussian approximation to $p(f(\theta))$.

Suppose the distribution of θ can be approximated by a normal distribution with mean $\hat{\theta}$ and covariance matrix $\mathbf{C}$. Let f be a function defined on an open subset of a d dimensional space taking values on r dimensional space and let f be differentiable at θ. The approximating distribution to $f(\theta)$ is the normal distribution with mean $f(\hat{\theta})$ and a covariance matrix:

$$\mathbf{C}_f = \nabla f^{\mathbf{T}} \, \mathbf{C} \, \nabla f \tag{3.26}$$

This approximation, which is discussed by Tanner [126], is variously called the δ method or the propagation of errors.

3.2 Optimization

The location of the dominant mode of a posterior density is the subject of considerable research [111]. Classical non-Bayesian statistics carry out inference on the basis of the likelihood function. More generally, one is interested in inverse theory, whose goal is to find physical models that fit the data optimally according to some prescribed criterion. In the Bayesian scheme the goal of *maximum a posteriori* estimation is to maximize the posterior density, or likelihood times prior, as a means of incorporating prior information missing from the likelihood scheme. More general Bayesian inference involving numerical integration of the posterior density also depends critically on the ability to locate the posterior mode. One example that we have seen already is the normal approximation to the posterior density. Many more examples will be presented later.

Optimization is usually stated in terms of minimizing a *cost function*[3]. In our case, we choose to minimize the negative of the logarithm of the posterior density rather than minus the posterior density itself. There are two definite advantages in working with log densities:

1. The density itself is a function whose value is subject to extreme variation (which can cause numerical problems on digital computers).

2. In the region of the mode the log density is approximately equal to a quadratic form, which is the form that is most suitable for many optimization techniques.

In a very limited number of cases the extrema (maxima and minima) may be located by differentiating the cost function, equating the derivative to zero and solving for the extrema in closed form. The next stage is classifying the extrema as either maxima, minima or saddlepoints using higher order derivatives. In the vast majority of cases this approach is not feasible and one must therefore resort to some numerical approach to optimization.

[3]It is important to distinguish between the general meaning of the term used in optimization and the very specific meaning used in decision theory and classification to define conditional risk and Bayes risk.

The likelihood is extremely expensive to compute. Hence, the chief factor that determines the effectiveness of an optimization algorithm is how few computations of the likelihood are required. Computational overheads within the algorithm that do not involve computing the likelihood usually constitute an insignificant proportion of the total computational effort.

Parameterization is a very important issue in optimization for two reasons:

1. If the parameterization is well chosen then it can make a multimodal problem into a unimodal one [111].

2. It is also convenient to map true parameter space onto a corresponding open space with parameters taking values along the real axis. This means that unconstrained optimization algorithms may be used, which simplifies optimization algorithms enormously.

We distinguish between those techniques that are dependent on the cost function being unimodal (*Local algorithms*) and those that are not (*Global algorithms*).

3.2.1 Local algorithms

Local optimization is the process whereby, starting from some point in parameter space, one can minimize the value of the cost function by taking a succession of downhill steps. If the cost function is multimodal then more likely than not this procedure will yield a suboptimal local minimum. Each step will reduce the value of the cost function. Our aim is to choose a method that accomplishes the minimization in as few function evaluations as possible.

There are numerous means for locating local minima. The usual approach is to minimize the cost function along n independent directions in turn (n line minimizations), repeating the process until the minimum is found. The simplest such approach is the *alternating variables* [34] method. The objective function is minimized along each parameter axis in turn, keeping the other parameters fixed; this procedure is repeated until a local minimum is found.

A better approach is to follow the line given by the gradient of the function to a local minimum, and starting at the new minimum take the gradient (the direction of which is always orthogonal to the direction of the previous gradient) and follow the line to the next minimum. This method is known as *steepest descent* [34, 95].

Neither alternating variables nor steepest descent is very good. This is especially true [34, 95] when the contours of the cost function are narrow and skewed at an angle to the parameter space as shown in figure 3.2.

Quadratic algorithms

There is a large family of optimization methods that operate on a local quadratic approximation to the cost function, the essential idea being that the Taylor expansion, unlike the cost function itself, is trivial to minimize. The local Taylor expansion is updated and minimized during each iteration of the optimization routine. By repeating this process (hopefully) the routine will converge to a local minimum.

The local quadratic approximation to a function $f(x+\delta)$ in the region about a point x in parameter space is:

$$f(x+\delta) \approx f(x) + \delta^{\mathbf{T}} \nabla f(x) + \frac{1}{2} \delta^{\mathbf{T}} \nabla\nabla^{\mathbf{T}} f(x)\, \delta \tag{3.27}$$

The right hand side is a quadratic approximation in δ.

The key technique that we have at our disposal is to *orthogonalize* the parameter space. By orthogonalizing we mean that we shall apply a change of basis to the parameter space in order to make the cost function easier to work with. Let δ be a parameter vector in the *old* parameter space and let $\mathbf{e}$ be the corresponding vector in the *new* parameter space. The two vectors are related by a linear transformation of the form

$$\mathbf{e} = \mathbf{T}\,\delta \tag{3.28}$$

where $\mathbf{e}$ is chosen such that

$$\mathbf{e}^{\mathbf{T}}\mathbf{e} = \delta^{\mathbf{T}}\,\mathbf{T}^{\mathbf{T}}\mathbf{T}\,\delta = \delta^{\mathbf{T}} \left[\nabla\nabla^{\mathbf{T}} f(x)\right] \delta \tag{3.29}$$

Conjugate directions

The method of conjugate directions is simply the implementation of the alternating variables method in orthogonal parameter space. The direction set $\{\mathbf{e_i} : 1 \le i \le d\}$ is orthonormal if:

$$\mathbf{e}_i^{\mathbf{T}}\mathbf{e}_j = \begin{cases} 1 & i = j \\ 0 & i \neq j \end{cases} \tag{3.30}$$

This is equivalent to

$$\delta_i^{\mathbf{T}} \left[\nabla\nabla^{\mathbf{T}} f(x)\right] \delta_j = \begin{cases} 1 & i = j \\ 0 & i \neq j \end{cases} \tag{3.31}$$

The direction set $\{\delta_{\mathbf{i}} : 1 \le i \le d\}$ are said to be *conjugate directions* [95].

Powell's contribution [95] was to derive a method for determining a set of conjugate directions using only function evaluations and line minimizations without knowledge of the Hessian matrix. Powell's method requires order $\mathcal{O}(d^2)$ function evaluations which is rather high but, as Press *et al* [95] explain, this is exactly what one should expect from any algorithm that uses function evaluations only.

Gradient based algorithms

The number of function evaluations required may be considerably reduced to order $\mathcal{O}(d)$ [95] by the use of gradient information.

Two families of gradient based algorithms may be used:

Variable metric (VM): This approach uses Newton-Raphson iterations to find a root of $\nabla f(x)$ and hence minimize the cost function $f(x)$.

Let us consider a linear Taylor expansion to $\nabla f(x)$

$$\nabla f(\mathbf{x} + \delta) = \nabla f(\mathbf{x}) + \nabla \nabla^{\mathbf{T}} f(x)\, \delta \tag{3.32}$$

If $\mathbf{x} + \delta$ is the location of the minimum along the direction of x then we have

$$\delta = -\left(\nabla \nabla^{\mathbf{T}} f(\mathbf{x})\right)^{-1} \nabla f(\mathbf{x}) \tag{3.33}$$

It is not possible to compute the step size δ without knowledge of the inverse Hessian matrix. The VM algorithm begins with a *positive definite* initial estimate for the inverse Hessian matrix; the identity matrix for instance. The estimate is updated using values for the gradient and the position of minima according to a rule which means that the current estimate for the Hessian matrix will converge to the true Hessian matrix:

$$\lim_{i \to \infty} H_i = \left(\nabla \nabla^{\mathbf{T}} f(x)\right)^{-1} \tag{3.34}$$

It is not critical to obtain the Hessian matrix exactly in order to reduce the value of the cost function at each iteration. In fact any *positive definite* matrix will suffice. Indeed, there are cases where the true Hessian matrix (local value away from the mode) is not positive definite and it would be counter-productive to use it.

The two different forms [95] of the VM algorithm are the Davidon-Fletcher-Powell (DFP) algorithm and the Broyden-Fletcher-Goldfarb-Shanno (BFGS) algorithm. The two algorithms differ only in the manner in which the approximation to the inverse Hessian matrix is computed. Press *et al* [95] hold that, in terms of robustness, BFGS is the superior of the two algorithms.

Conjugate gradients: The conjugate gradient method is similar in principle to the method of conjugate directions. Press *et al* [95] describe how gradient information may be employed to construct a set of conjugate directions along which one may minimize the cost function. This is carried out without having to compute the Hessian matrix. In addition, gradient information can be used in the line minimization itself by using the Newton-Raphson method.

There are two variants of the conjugate gradient method. The more straightforward is known as the Fletcher-Reeve algorithm [95]. A slight modification may be made to the Fletcher-Reeve algorithm to make it more robust to non-quadratic local minima. This yields the Polak-Ribière [93, 95] algorithm.

Non-quadratic algorithms

Many algorithms also exist which do not use quadratic approximations to the cost function. The most notable among these is the *simplex algorithm* of Nelder and Mead [83, 95]. The basis of this algorithm is that one constructs a d dimensional simplex in parameter space consisting of $d+1$ vertices such that the enclosed volume of the simplex is non-zero. On each iteration the vertex with the highest value of cost function is replaced with a new vertex whose position is computed by projecting through the vertex with lowest cost function. Other operations that may be carried out to vary the position of the simplex include reflections, contractions and expansions [83, 95]. This algorithm is not very efficient in terms of the number of function evaluations that are required although it is usually very robust [95] in most applications.

The Hooke and Jeeves algorithm [19, 51] has been applied to optimization for Bayesian inference [19, 32]. This algorithm is based on the alternating variables method described above, but it includes heuristics to speed up the progress of the iteration along long narrow valleys in the cost function. Fletcher [34] holds that the Hooke and Jeeves algorithm is far inferior to Powell's method outlined above.

The expectation-maximization (EM) algorithm [29, 126] is designed for the solution of missing data problems using maximum likelihood, but it has also been used to solve problems with no missing data [128] by the clever use of dummy variables to represent the missing data. The EM algorithm will be applied to interpolate missing samples of autoregressive data in chapter 6.

Memory requirements

The optimization of functions of a very large number of parameters is often difficult because of memory requirements. The VM algorithm requires $\mathcal{O}(d^2)$ units of storage whereas the conjugate gradients algorithm's memory requirements are in the order $\mathcal{O}(d)$. Lasenby and Fitzgerald [69] present a scheme, similar to conjugate gradients, that is suitable for the optimization of large scale problems.

3.2.2 Global algorithms

Global optimization locating the global minimum of the cost function regardless of the starting point in parameter space. Global optimization methods are designed to overcome the multimodality of the cost function without incurring the great expense of an exhaustive search.

For most Bayesian applications the data is noisy which means that a very precise answer is not usually required. It is sufficient to get near to the global optimum. Two candidate optimization methods are *simulated annealing* and *genetic algorithms.*

In general, global optimization techniques search the entire parameter space while concentrating on the region around where the last good minimum of the cost function was found. The exploration of parameter space is carried out stochastically using pseudo-random variates. This is as much as simulated annealing and genetic algorithms have in common.

Genetic algorithms

The term "genetic algorithm" [42] is applied to an optimization method based on the Darwinian theory of the survival of the fittest. A large population of trial solutions is proposed with each individual trial solution being encoded in the form of a binary bit string. The fitness of a candidate solution from the population depends on how well it minimizes the cost function.

The next population is bred from the current population using two heuristics:

Breeding: Two parents are selected from the population according to their fitness. A new bit string is constructed drawing one bit at a time from each of the parents' bit strings in the same way as genes are selected at random during natural selection.

Mutation: Very occasionally a small number of bits are flipped at random.

Repetition of this process can lead to some very good solutions amongst the population, in the sense that a near optimal solution can be obtained.

Simulated annealing

Simulated annealing is an optimization method that is based on an analogy with the behaviour of hot materials as they are cooled. The classic example is the slow cooling of very hot metals: a slowly cooled metal has much larger average grain size than a metal that is cooled quickly. This process of cooling a metal slowly to achieve this aim is known as *annealing*. The energy of a regular configuration of atoms or molecules is lower than a disordered configuration. In other words, the slowly cooled metal achieves a lower energy state than the quickly cooled metal.

The probability density of the states in a solid, ignoring quantum effects, may be assumed to agree with the canonical[4] distribution

$$p_B(E) = \frac{1}{Z} \exp\left[-\frac{E}{kT}\right] \tag{3.35}$$

where E is the energy associated with a given state, T is the absolute temperature and k is Boltzmann's constant. The constant Z is a normalization constant known as the partition function. Note that the state of highest probability density is the state of minimum energy.

The concept of energy and state are quite arbitrary. For our purposes we can equate state with a position in parameter space and energy with the corresponding value of the cost function.

Simulated annealing is a direct analog of true annealing; starting at a sufficiently high temperature so all states are readily accessible, the temperature is gradually lowered towards absolute zero. In theory, all states are equiprobable at infinite temperature, while at the other extreme only the global minimum is possible at zero temperature. The principle behind simulated annealing is that random states are drawn from the canonical distribution in equation 3.35, while slowly reducing the temperature. By analogy with the microcanonical states of the metal, we expect that the energy, which is the value of the cost function to be minimized, will be reduced to a relatively low value.

The task of drawing a random sample from the canonical distribution may be achieved in a number of different ways. If the distribution is of a particularly simple form (i.e. Gaussian, inverse gamma etc.) then it may be carried out efficiently using some simple method. In general however, one may have to resort to sophisticated sampling techniques such as the Metropolis algorithm or Hybrid Monte Carlo as described in the next chapter.

The real power of a sampling based approach is that the energy does not always decrease but can sometimes actually increase, depending on the outcome of sampling from the canonical distribution in equation 3.35. This approach is a fundamental departure from the "greedy" local optimization algorithms such as BFGS or DFP [95] mentioned earlier. The random search procedure enables one to escape from local minima of the energy function. There is a general result [36, 82] which shows that if the temperature is held fixed then the Metropolis random walk or Hybrid Monte Carlo trajectories will, given an infinite amount of time, visit *every* state with a relative frequency given by the canonical probability density in equation 3.35.

The crucial factor determining the success of simulated annealing is the choice of *annealing schedule*. It can be shown [36, 53, 82] that certain an-

[4]This is also known as either the Boltzmann distribution or the Gibbs distribution.

nealing schedules [5] will always lead to the global minimum of the energy function. We will discuss annealing schedules and simulated annealing in far greater detail in section 4.7.

Genetic algorithms versus simulated annealing

Both simulated annealing and genetic algorithms have their own groups of devotees. Scales, Smith and Fisher [111] compare the performance of both approaches on seismic data and find that genetic algorithms give better results, justifying their results in terms of the memory of previous states embedded in the genes, and suggesting that somehow the GA has evolved and learned about the cost function. On the other hand, Ingber and Rosen [53] compare their own "very fast simulated reannealing" (VFSR) approach with genetic algorithms and find that VFSR is far superior.

Hence, the difference in performance between the two methods is difficult to judge. Press *et al* [95] refer to the "tender loving care factor", which means that even the most objective author will always know how to adjust his particular algorithm to obtain the best performance. This of course gives it an unfair advantage over other techniques and makes it difficult to assess the relative performance of new techniques. In addition, there will always be problems where either method will give better results.

In this book, we favour simulated annealing for one very good reason: one of the most effective means of carrying out multidimensional numerical integration is to use Monte Carlo integration, which involves sampling from the posterior density (or an approximation to it). Once advanced sampling techniques are in place then it becomes a relatively trivial matter to implement simulated annealing, rather than genetic algorithms, the only work required being the selection of an annealing schedule appropriate to the given optimization problem. This point will be illustrated by example in the coming chapters.

3.2.3 Concluding remarks

The posterior density is often a difficult function to maximize because it can be multimodal and is subject to extreme variation, and the parameter space may be composed of both continuous and discrete parameter subspaces. A number of common approaches to optimization have been discussed in order to propose a suitable solution.

In the case of continuous parameter spaces, if the sample size is large then asymptotic normality will apply. Asymptotic normality does not necessarily imply unimodal behaviour. We have seen a one dimensional example in

[5]This schedules are usually impractical because it takes an infinite amount of time to guarantee convergence, and an impossibly long time to attain a high probability of obtining a good solution.

chapter 2. Figure 2.2 shows the marginal density of a single frequency, which appears to be almost Gaussian. Figure 2.3 shows the corresponding log density which is multimodal.

Generalizing the single frequency example to other Bayesian probability densities, it is obvious that a global optimization algorithm (such as simulated annealing) is required in the initial stages to locate the dominant mode of the posterior density. Thereafter, a local algorithm such as BFGS can be used to "polish" up the results. Many local optimization algorithms are well suited to optimizing the log density in the region of the mode because of its quadratic nature.

3.3 Integration

This book is mainly concerned with the problem of approximating integrals for the purpose of carrying out Bayesian inference.

Press *et al* [95] note that "integrals of functions of several variables, over regions with dimension greater than one, are *not easy*." Although this is true in general there are certain provisos which need to be added when referring to Bayesian integrals. Compared to integration problems in general those arising in Bayesian statistics have distinctive characteristics; namely, that they involve integrands that are often smooth and have a single dominant peak in some region in parameter space. In these cases, it is obviously desirable that we employ computational methods that make full use of the information at hand about the integrand in order to choose a strategy that is as efficient as possible.

Three different strategies are often used for computing Bayesian integrals:

1. *Product Rule Quadrature*
2. *Asymptotic Approximation*
3. *Monte Carlo Integration*

These methods are applied to the computation of three different quantities:

Evidence: Evidence is a real number whose value is used as a merit index for comparing the performance of different models at describing the data. We can either calculate *absolute* evidence, which means the normalizing factor for the product of likelihood and prior, or calculate *relative evidence*, which is the ratio of the evidence of one model compared to another.

Marginal density: A marginal density is a function of one or more parameters that is used to compare the plausibility of different parameter values, and is as such used for parameter estimation. The shape of the marginal density is more important than its size and, in fact, we almost always dispense with the computation of the normalization factor. Because the marginal density is a curve we must compute enough sample points for it to be possible to make an accurate plot of the density. Typically, about thirty sample points are required for a good representation of a one dimensional marginal density. This means evaluating no less than thirty integrals!

Moments and expectations: Computing moments and expectations of functions is not significantly different to computing marginal densities or evidence, depending on whether the result is in the form of a curve or a single real number. However, unlike marginalization, it is important not to discard normalization constants.

The difference in the kind of result required leads to considerable differences between the techniques developed to evaluate these three quantities.

There is a lot that can be done to make a Bayesian integral *easier* to do than a typical multidimensional integration. The most powerful weapon in the armoury is that of parameterization. Orthogonalizing transformations have the effect of making the integral more tractable, by producing an integrand that is approximately separable.

The most important result from our point of view is asymptotic normality. In the large sample data scenario normal approximations to the posterior density can be very accurate indeed. The density may be normal to a good approximation [131] even for moderate amounts of data. More importantly from our point of view, it is a good starting point for a full Bayesian analysis, because it gives some idea of the scale and orientation of the posterior density, and thus leads to much more efficient sampling. This point is discussed in more detail later in the chapter.

In addition, it is noted by Tierney and Kadane [131] that the results obtained need not be extremely accurate because the statistical uncertainty due to the noise corrupting the data is likely to be quite large.

3.4 Numerical quadrature

In this section we will discuss numerical quadrature and its application to Bayesian integrals. By numerical quadrature, we mean any rule similar to Simpson's rule which samples the integrand at prescribed points on a grid according to a given formula. These methods may be used for computing either evidence or marginal densities as well as posterior moments and expectations. Consider an integral of the form:

$$g(s) = \int_{-\infty}^{\infty} f(t, s)\, dt \tag{3.36}$$

Let us also assume that there is sufficient data for normality to begin to assert itself, but that the number of data are such that the normal approximation is not a particularly good approximation.

Naylor [80], Naylor and Smith [81] and Smith [119] propose the following approximation to the integrand:

$$f(t, s) \approx \exp\left[-\frac{(t-\mu)^2}{2\sigma^2}\right] p(t, s) \tag{3.37}$$

where $p(t, s)$ is a polynomial in t of order $2M$.

The product of a Gaussian and a polynomial may be integrated *efficiently* by means of Gauss-Hermite[6] quadrature:

$$\int_{-\infty}^{\infty} f(t, s)\, dt \approx \sum_{i=1}^{M} w_i\, f(t_i, s) \exp\left[{x_i}^2\right] \tag{3.38}$$

where the weights w_i and the abscissae t_i are given by the formulae:

$$w_i = 2^{M-1} M! \sqrt{\pi}\, H^2_{M-1}(x_i) \tag{3.39}$$

$$t_i = \sqrt{2}\,\sigma\, x_i + \mu \tag{3.40}$$

and where x_i is the ith root of the Hermite polynomial $H_M(x)$. Tables of x_i and $w_i \exp\left[{x_i}^2\right]$ are listed by Davis and Polonsky [27].

The main feature of Gaussian quadrature is that, unlike Simpson's rule, the samples are non-uniformly spaced. The samples in equation 3.38 are deliberately positioned at the roots of the Mth Hermite polynomial, which means that the formula will exactly integrate the product of a Gaussian and a polynomial of order $2M$. Uniform sampling (using a Newton-Cotes formula) will only allow exact integration of the product of a Gaussian and a polynomial of order M. A mathematical derivation [6] of this result uses the properties of orthogonal polynomials. Good reviews of Gaussian quadrature (including Gauss-Hermite, Gauss-Laguerre and Gauss-Legendre) are given by Arfken [6], Davis and Rabinowitz [28] and Stroud [124].

3.4.1 *Multiple integrals*

Multiple integrals may be computed using "nested loops" of univariate integrals. A product rule will express a multiple integral in the following

[6]The term "Gaussian" quadrature has nothing to do with the fact that in this case the integrand is approximately Gaussian. It is in fact the term given to quadrature that samples the integrand at the roots of orthogonal polynomials as a means of increasing the order of accuracy of approximation to the integral.

form:

$$\int\int\int\cdots\int f(t_1, t_2, t_3, \ldots t_N,)\, dt_1\, dt_2\, dt_3 \ldots dt_N$$
$$= \sum_{i=1}^{M_1} w_i \sum_{j=1}^{M_2} w_j \sum_{k=1}^{M_3} w_k \cdots \sum_{l=1}^{M_N} w_l\, g(x_1^{(i)}, x_2^{(j)}, x_3^{(k)}, \ldots x_N^{(l)}) \quad (3.41)$$

This requires the user to specify a grid. By a "grid" we mean choosing a coordinate basis and sampling at fixed points along each coordinate. From the point of view of numerical integration the most efficient grid to use is an orthonormal basis as described by Naylor and Smith [81].

Determining a suitable orthonormal basis presents some difficulty. Naylor and Smith [80] suggest a rather clever scheme to determine just such an orthogonal basis. Using the covariance matrix at the mode of the joint posterior (estimated using equation 3.6) as a good starting point, one can compute the first and second moments of the distribution. This gives rise to a new estimate for the mean and the covariance matrix which may be used to develop a new grid. The process is iterated until convergence. This scheme, though ingenious, does not seem to work well in practice, the essential problem being that it is a non-linear iterative map that does not always converge.

Function evaluations

The number of function evaluations required to compute an integral equals the number of points on the grid. This brings us to the essential problem of using Gaussian quadrature methods, which is how the number of computations required to evaluate a multidimensional integral increases with dimensionality. Consider for example a three dimensional integral being computed using a five point rule (i.e. $M = 5$ in equation 3.38). This would require $5^3 = 125$ function evaluations. A six dimensional integral would require $5^6 = 15,625$ function evaluations. This is a rather frightening rate of increase, with the number of computations growing exponentially in the number of dimensions. Naylor and Smith [80] present tables to illustrate this phenomenon.

The number of computations may be reduced considerably by exploiting the spherical symmetry of the orthogonalized parameter set. The coordinate basis is transformed from a Cartesian coordinate system to a spherical one. Gauss-Hermite quadrature is then replaced by Gauss-Laguerre quadrature in the radial component and Gauss-Legendre in the angular component. The growth in the number of computations is now of the order 2^d instead of M^d (where d is the dimensionality). As noted by Smith [119], this allows one to use Gaussian quadrature to handle higher dimensional problems, although the fundamental problem of exponential growth in the number of function evaluations still remains. The disadvantage of using a spherical grid rather than a Cartesian grid is that sample points from the marginal density are not available as before[7].

Accuracy

Wolpert [139] identifies another serious deficiency in using Gaussian quadrature, namely its lack of accuracy. The integral in equation 3.38 assumes that the integrand may be expressed as the product of a Gaussian and a polynomial. Not all probability densities can be described in this form. This means that systematic errors are introduced by using such a formula. The problem becomes severe when computing multiple integrals so, as Wolpert notes, not only do the integrals take a long time to compute but the results that are obtained for high dimensional problems can easily be completely meaningless.

From the above discussion it is clear that Gaussian quadrature is simply not an effective means of carrying out numerical Bayesian inference in large scale problems (i.e. dimensionality greater than ten), because the number of computations becomes prohibitive and the accuracy is often poor. Not surprisingly, alternative approaches to numerical integration for Bayesian applications that give better accuracy for far less computation have been devised.

3.5 Asymptotic approximations

Bayesian integrals tend to depend very strongly on the sample size. In this section, we shall exploit this dependence to compare the given integral with a simpler one that can be evaluated exactly, yet closely approximates the original.

In most cases the integrand can be expressed in the form of a Gaussian multiplied by a Taylor series of the sample size. If the approximating series is a uniformly good approximation to the integrand (for all values of the variable of integration) then the series may be integrated term by term to give a good approximation to the integral.

Asymptotic approximations to a density may be written in the form:

$$f(x) = g(x)\left(1 + \frac{a_1(x)}{N^{1/2}} + \frac{a_2(x)}{N} + \frac{a_3(x)}{N^{3/2}} + \mathcal{O}\left(N^{-2}\right)\right) \tag{3.42}$$

where N is the sample size. The term in $g(x)$ is Gaussian and is often exactly the same as the normal approximation discussed earlier in section 3.1.

---

[7] Although $M = 5$ sample points from a marginal density tends to be a little inadequate!

3.5.1 The saddlepoint approximation and Edgeworth series

The saddlepoint approximation (or method of steepest descent) is common throughout physics [6] and engineering [25]. Daniels [26] first introduced the saddlepoint approximation to statistical applications. The saddlepoint approximation along with the related Edgeworth expansion have since been applied by many authors. For example, Barndorff-Nielsen and Cox [8] use the saddlepoint approximation and Edgeworth expansion for approximating the density of the sum of independent random variables.

The saddlepoint approximation to a probability density function is computed by inverting its Laplace transform (the moment generating function) by contour integration. The Edgeworth expansion is similar in principle except that it expands the logarithm of the moment generating function.

Two major points about the approximation in equation 3.42 are:

1. The approximation is Gaussian with a multiplicative factor that is a polynomial in $N^{-1/2}$. The terms of the polynomial decrease uniformly with increasing N so the density will converge to $g(x)$.

2. The coefficients of the polynomial are themselves polynomials in x. In the case of the saddlepoint approximation the polynomials will consist of simple powers of x. In Edgeworth expansions the coefficients take the form of Hermite polynomials scaled by a factor involving cumulants of the distribution.

Barndorff-Nielsen and Cox [8] have also worked out multivariate approximations. These are potentially useful for computing marginal densities since the analytic form of Gaussian times polynomial may be readily integrated as a gamma integral.

The advantage of using saddlepoint and Edgeworth expansions is that they are in closed form. There are also some major disadvantages:

- The approximation to thick tailed densities can be poor in the tails.
- The approximation is only exact in the limit as the sample size approaches infinity. In practice it can be difficult to judge the accuracy of the approximation for moderate sample size.
- These approximations tend to become rather complicated and cumbersome to work with.

These asymptotic approximations have by and large been superceded by the simpler Laplace approximation.

3.5.2 The Laplace approximation

In this section we discuss various approximations that can be made to compute expectations and marginal densities. As before in the case of the normal approximation, the philosophy will be to replace difficult integrations with conditional maximizations and numerical differentiation.

3.5.3 Moments and expectations

Suppose $h(\theta)$ is a smooth, bounded unimodal function with a maximum at $\hat{\theta}$ where θ is a scalar. By Laplace's method, one can expand the integral

$$I = \int f(\theta) \exp\left[-n\, h(\theta)\right] d\theta \approx f(\hat{\theta}) \sqrt{\frac{2\,\pi}{n}}\, \sigma \exp\left[-n\, h(\hat{\theta})\right] \tag{3.43}$$

where

$$\sigma = \left[\left. \frac{\partial^2 h}{\partial \theta^2} \right|_{\theta=\hat{\theta}} \right]^{-1/2} \tag{3.44}$$

and n is the sample size.

Mosteller and Wallace [79] obtain this result by expanding $h(\theta)$ about $\hat{\theta}$ to give

$$\hat{I} = \int f(\theta) \exp\left[-n \left(h(\hat{\theta}) + \frac{\left(\theta - \hat{\theta}\right)^2}{2} \left. \frac{\partial^2 h}{\partial \theta^2} \right|_{\theta=\hat{\theta}} \right) \right] d\theta \tag{3.45}$$

The integral in equation 3.45 is readily computed to give the integral in equation 3.43.

It is intuitively clear that if $\exp\left[-n\, h(\theta)\right]$ is very peaked about $\hat{\theta}$ then the integral will be dominated by the behaviour of the integral about $\hat{\theta}$. More formally, one can show [131] that

$$I = \hat{I} \left(1 + \mathcal{O}\left(\frac{1}{n}\right) \right) \tag{3.46}$$

where n is the sample size.

To calculate the moments of a posterior distribution of an expectation we need to evaluate expressions of the form:

$$E\left(g(\theta)\right) = \frac{\int g(\theta) \exp\left[-n\, h(\theta)\right] d\theta}{\int \exp\left[-n\, h(\theta)\right] d\theta} \tag{3.47}$$

Kass, Tierney and Kadane [60], Tierney and Kadane [131] and Tierney, Kass and Kadane [133] give two approximations for $E\left(g(\theta)\right)$:

1. Let $f = g$ in equation 3.43 to evaluate the numerator of expression 3.47. Also, let $f = 1$ in equation 3.43 to evaluate the denominator of expression 3.47. This gives:

$$E\left(g(\theta)\right) = g(\hat{\theta}) \left(1 + \mathcal{O}\left(\frac{1}{n}\right) \right) \tag{3.48}$$

We shall refer to this as the *profile* approximation.

2. The second order approximation to the expectation is given by:

$$E(g(\theta)) = \frac{\sigma^* \exp[-n\,h^*(\theta^*)]}{\sigma \exp[-n\,h(\theta)]}\left(1 + \mathcal{O}\left(\frac{1}{n^2}\right)\right) \tag{3.49}$$

where the significance of the terms are as follows. Firstly we have

$$-\,n\,h^*(\theta) = -n\,h(\theta) + \log[g(\theta)] \tag{3.50}$$

and also

$$\begin{aligned} \sigma^* &= \sqrt{\det \Sigma^*} \\ \sigma &= \sqrt{\det \Sigma} \end{aligned}$$

and

$$\begin{aligned} \Sigma^* &= \left[\left. \frac{\partial^2 h^*}{\partial \theta^2} \right|_{\theta^*} \right] \\ \Sigma &= \left[\left. \frac{\partial^2 h}{\partial \theta^2} \right|_{\hat{\theta}} \right] \end{aligned}$$

with $\hat{\theta}$ being the supremum of $-n\,h(\theta)$ and θ^* being the supremum of $-n\,h^*(\theta)$.

The obvious disadvantage with this approximation is that $g(\theta)$ must be a positive function. Tierney, Kass and Kadane [134] have proposed alternative approximations which overcome this difficulty.

Lindley [73, 119] earlier proposed an alternative approximation which requires the evaluation of third order derivatives.

Note that equation 3.47 is the ratio of two very similar integrals. Tierney and Kadane [131] describe how this results in the high order of accuracy in the approximation because of the normalization terms cancelling.

3.5.4 Marginalization

Marginalization is a slightly easier problem than computing expectations because the aim of the exercise is to obtain a function that is, as far as possible, proportional to the true marginal density. Normalization constants are not a relevant issue.

Let us partition the p dimensional space Θ. The $p \times 1$ vector θ may be written as (θ_1, θ_2) where θ_1 is a scalar and θ_2 is $(p-1) \times 1$. Let $\hat{\theta}_2(\theta_1)$ maximize the posterior $p(\theta \mid \mathbf{d})$ for a given value of θ_1 and let

$$\hat{\Sigma}^* = \left[\left. \frac{\partial^2 \log p\,(\theta_1, \theta_2)}{\partial \theta_2^2} \right|_{\theta_2 = \hat{\theta}_2} \right]^{-1} \tag{3.51}$$

which is an inverse Hessian matrix of size $(p-1) \times (p-1)$.

Tierney and Kadane [131] approximate $p(\theta_1 \mid \mathbf{d})$ with the following two approximations:

$$\textbf{Laplace:} \quad p(\theta_1 \mid \mathbf{d}) \propto p\left(\theta_1, \hat{\theta}_2 \mid \mathbf{d}\right) \left(\det \hat{\Sigma^*}\right)^{1/2} \left(1 + \mathcal{O}\left(\frac{1}{n}\right)\right) \tag{3.52}$$

$$\textbf{Profile:} \quad p(\theta_1 \mid \mathbf{d}) \propto p\left(\theta_1, \hat{\theta}_2 \mid \mathbf{d}\right) \left(1 + \mathcal{O}\left(\frac{1}{n^{1/2}}\right)\right) \tag{3.53}$$

The profile posterior is a simple statement of the fact that the integral is dominated by the conditional maximum value of the integrand. The Laplace approximation may be viewed as a simple modification of the profile posterior, in that it takes into account the "width" of the integrand about the conditional maximum as well as its height. The determinant term is proportional to the hypervolume under a Gaussian approximation to the conditional density.

Tierney and Kadane [131] present the Laplace approximation for a marginal density with an estimate of the normalization constant included:

$$p(\theta_1 \mid \mathbf{d}) \approx \frac{1}{\sqrt{2\,\pi}} \frac{p\left(\theta_1, \hat{\theta}_2 \mid \mathbf{d}\right) \left(\det \hat{\Sigma^*}\right)^{1/2}}{p\left(\hat{\theta}_1, \hat{\theta}_2 \mid \mathbf{d}\right) \left(\det \hat{\Sigma}\right)^{1/2}} \left(1 + \mathcal{O}\left(\frac{1}{n}\right)\right) \tag{3.54}$$

where

$$\hat{\Sigma} = \left[\left. \frac{\partial^2 \log p\,(\theta_1, \theta_2)}{\partial \theta^2} \right|_{\theta=\hat{\theta}} \right]^{-1} \tag{3.55}$$

which is an inverse Hessian matrix of size $p \times p$.

The approximation in equation 3.54 is only accurate to order $\mathcal{O}(n^{-1})$ because of the error in the estimated normalization constant. The accuracy of the approximation is increased to order $\mathcal{O}(n^{-3/2})$ [133] if expression 3.52 is integrated to obtain the normalization constant *exactly*. This is a straightforward one dimensional integral. If one has evaluated the marginal density at enough points to plot the curve then it follows that there should be enough points to evaluate the normalization constant accurately. For this reason, it is better to work with equation 3.52 directly and ignore the estimation of the normalization constant.

Non-linear parameterizations

Often it is required to approximate the marginal posterior density $p\,(\eta \mid \mathbf{d})$ of a parameter of interest where

$$\eta = g\,(\theta) \qquad\qquad \eta \in \Re \tag{3.56}$$

and g is a real continuous function defined on $\Re^p$.

In most situations it is virtually impossible to perform the tedious $p-1$ dimensional numerical integrations for computing the exact density

$$p(\eta \mid \mathbf{d}) = \int_D p(\theta \mid \mathbf{d})\, d\theta \tag{3.57}$$

where D denotes the region

$$D = D(\eta, \gamma) = \{\theta : \eta \leq g(\theta) \leq \eta + \gamma\} \tag{3.58}$$

Tierney, Kass and Kadane [132] propose the following approximation:

$$p(\eta \mid \mathbf{d}) = \frac{p_M(\eta \mid \mathbf{d})}{\sqrt{|\mathbf{R}_\eta|\, \mathbf{b}_\eta^{\mathbf{T}}\, \mathbf{R}_\eta^{-1}\, \mathbf{b}_\eta}} \tag{3.59}$$

where

$$p_M(\eta \mid \mathbf{d}) = \sup_{\theta : g(\theta) = \eta} p_M(\theta \mid \mathbf{d}) \tag{3.60}$$

$$\mathbf{b}_\eta = \left. \frac{\partial g(\theta)}{\partial \theta} \right|_{\theta = \theta_\eta} \tag{3.61}$$

$$\mathbf{R}_\eta = \left. \frac{\partial^2 \log p(\theta \mid \mathbf{d})}{\partial \theta^2} \right|_{\theta = \theta_\eta} \tag{3.62}$$

where θ_η conditionally maximizes the posterior density with respect to θ for each η.

Leonard, Hsu and Tsui [71] note that expression 3.59 becomes singular if the first derivative of g approaches zero. They propose an alternative approximation which does not require evaluation of the derivative.

3.6 The Monte Carlo method

The principle behind the Monte Carlo method is quite simple. Suppose a quantity expressed in integral form has to be evaluated. The Monte Carlo approach is to specify that quantity as the mean (or some other moment[8] or statistic) of a random process.

As a simple example, consider the problem of estimating the variance of a zero mean Gaussian process whose probability density function is given by:

$$p(x) = \frac{1}{\sqrt{2\, \pi\, \sigma^2}} \exp\left[-\frac{x^2}{2\, \sigma^2}\right] \tag{3.63}$$

[8]Hammersley and Handscomb [47] present an example where the value for π is estimated from the *variance* of a random process.

The variance v is given by the integral:

$$v = \int_{-\infty}^{\infty} x^2 \, p(x) \, dx \tag{3.64}$$

In other words to calculate the variance one must take every value of x^2 and weight it with $p(x)$. A Monte Carlo approach to estimating the variance is as follows:

1. Simulate N random variates from a Gaussian distribution

$$x_i \leftarrow N\left(0, \sigma^2\right)$$

2. Compute the average

$$v \approx \frac{1}{N} \sum_{i=1}^{N} x_i^2$$

The weighting of the term in x^2 is implicit in the sampling because the x_i^2 are sampled from $p(x)$.

The generalization of this approach to more complex problems, such as the evaluation of numerical integrals, is not difficult. Consider the following integral:

$$I = \int_X f(x) \, dx \tag{3.65}$$

This integral may be expressed in the form:

$$I = \int_X \frac{f(x)}{g(x)} \, g(x) \, dx \tag{3.66}$$

where the function $g(x)$ is normalized:

$$\int_X g(x) \, dx = 1 \tag{3.67}$$

Using the simple example of equation 3.64 as a guide, we can estimate the integral in equation 3.65 by:

1. Simulate N random variates from a distribution $g(x)$

$$x_i \leftarrow g(x)$$

2. Compute the average

$$I = E_g\left(\frac{f(x)}{g(x)}\right) \approx \frac{1}{N} \sum_{i=1}^{N} \frac{f(x_i)}{g(x_i)} \tag{3.68}$$

This technique is known as "importance sampling" because $f(x)$ is being sampled non-uniformly with the density $g(x)$, thereby giving more "importance" to some values of $f(x)$ than others. The variance of the estimate in equation 3.68 depends on the sampling density $g(x)$. The exact expression for the variance is given by:

$$\operatorname{var}\left[E\left(\frac{f(x)}{g(x)}\right)\right] = \int_X \left(\frac{f(x)}{g(x)} - I\right)^2 g(x)\, dx \tag{3.69}$$

Note that the variance of the estimate is zero if:

$$g(x) = \frac{f(x)}{\int_X f(x)\, dx} \tag{3.70}$$

which is the case when we sample variates directly from the integrand $f(x)$. In general we may not sample directly from the integrand[9] because of the difficulties involved in generating random samples with a given distribution. In any case even if it is possible to generate random samples with the desired distribution, it will be impossible to evaluate g (x) since to do so would require knowledge of the normalization constant (the denominator of equation 3.70), which is tantamount to knowing the answer to the problem before one starts.

In fact in theory *almost any*[10] importance sampler can be used. The art of importance sampling is to choose a density $g(x)$ that resembles the integrand $f(x)$ as closely as possible. This is a non-trivial problem since we must be able to draw variates from the density $g(x)$ while at the same time having exact knowledge of the normalization factor.

Another important point to note is the slow rate of convergence for the estimate of the integral. It can be shown [139] that the variance of the estimate in equation 3.68 (using pseudo random variates) is of order $\mathcal{O}(N^{-1})$, where N is the number of function evaluations. The estimate of an integral does not improve considerably with an increased number of function evaluations. The degree to which the sampling density $g(x)$ approximates the integrand $f(x)$ is the principal factor determining the effectiveness of importance sampling.

A key point is that unlike the quadrature or asymptotic approximations described previously, the Monte Carlo approach exhibits *no systematic error*. The Monte Carlo estimates that we will discuss are unbiased[11]. Any error in the estimated integral is caused by random effects due to sampling. In addition the variance of the estimate can also be readily obtained using

[9] One may sample the integrand using Markov chains but the normalization constant is still unknown.

[10] One condition is that $g(x) > 0$ if $f(x) > 0$.

[11] This will not be the case for sampling importance resampling (SIR).

equation 3.69. This contrasts with quadrature and asymptotic approximation where very often there is no reliable estimate of the accuracy of the approximation.

The most important point is dimensionality. There is no *direct* dependence between computational requirements and dimensionality in Monte Carlo integration. Monte Carlo integration avoids the need for a grid as required by quadrature simply because it does not sample at prescribed points in parameter space. In addition, quadrature accumulates systematic errors as dimensionality increases. For large scale problems this error can be quite severe. Wolpert [139] notes the Monte Carlo integral is often therefore the most accurate approximation available.

3.7 The generation of random variates

In this section, we will give a very brief survey of the generation of random variates having a prescribed distribution. The aim ultimately is to provide random variates primarily for use in Monte Carlo integration but they may also be used for Monte Carlo simulation and for use in numerical optimization using simulated annealing or genetic algorithms.

We begin by stating an important fact that is often overlooked: the generation of random variates with a given density is *not easy*.

3.7.1 Uniform variates

All synthetic random variates start off life as uniform random variates. The foundation underpinning all stochastic simulations is the ability to generate uniform random variates over the range (0, 1].

The generation of uniform variates by computer is full of potential pitfalls with which one must be aware. It is generally assumed as an article of faith that the particular random number generator is adequate for the purpose for which it is being employed. Computer generated random variates are of course not truly "random" but are usually computed using a recurrence of the form:

$$x_{i+1} = (a\,x_i + b) \bmod m \tag{3.71}$$

This is the linear congruential generator. Suitably chosen values of a, b and m will lead to good "random" sequences with a period m that will pass almost all statistical tests of randomness.

Press *et al* [95] note some of the more unusual properties of such sequences. The most serious from our point of view is the presence of serial correlations. If the output sequence is segmented into k-vectors and plotted in k-dimensional space, the points will lie on at most $m^{1/k}$ hyperplanes. It is potentially disasterous for numerical integration and optimization if the regions of greatest magnitude of the posterior density are very narrow and therefore lie between hyperplanes.

3.7.2 Non-uniform variates

The art of generating non-uniformly distributed variates is fundamental to Monte Carlo integration. Ripley [104] and Devroye [30] give good reviews of current techniques. We now review the basic approaches.

Inverse cdf

Given a uniform variate one way to produce a non-uniform variate is to transform the variable using a suitable function. To generate a variate x_i with cumulative distribution function $G(x)$ the most straightforward method is to apply:

$$x_i = G^{-1}(u_i) \tag{3.72}$$

There is a one to one correspondence between the uniform variate u_i and the non-uniform variate x_i.

The inverse transform method is only useful for densities where the inverse cumulative distribution function may be expressed in closed form. For instance, Gaussian variates are not easy to generate in this way because of the difficulty of inverting:

$$G(t) = \frac{1}{\sqrt{2\pi\sigma^2}} \int_{-\infty}^{t} \exp\left[-\frac{x^2}{2\sigma^2}\right] dx$$

It is clear that approaches other than the inverse transform method will usually be required for simulating densities.

3.7.3 Transformation of variables

In the previous section, we discussed how to generate random variates with a given distribution by inverting the cumulative distribution function. This is in fact a special case of the fundamental transformation law [89, 90] of probability. Suppose we generate a random variate x and take some prescribed function $y(x)$ of it, then the probability distribution of $y(x)$ is related to the probability of x by

$$|p(y)\,dy| = |p(x)\,dx| \tag{3.73}$$

which implies

$$p(y) = p(x)\left|\frac{\partial x}{\partial y}\right| \tag{3.74}$$

Setting $p(x)\,dx$ to a uniform density gives the rule used in the previous section.

3.7.4 The rejection method

The rejection method is a flexible approach for the generation of random variates. Suppose it is desired to draw random variates from the density $f(x)$. If there exists a density $c(x)$ from which we can sample variates easily and where we have

$$k\,c(x) > f(x) \tag{3.75}$$

where k is a real positive constant then Press *et al* [95] illustrate how the following steps will generate samples from $f(x)$:

1. Generate a random variate $x_i \leftarrow c(x)$
2. Accept x_i with probability $f(x)/k\,c(x)$

Step 1 may involve generating a uniform random variate and inverting the cumulative distribution function as described earlier. Step 2 may be implemented by using a random uniform variate; if the random variate is less than the ratio then the variate x_i is accepted. The efficiency of this scheme, which is the ratio of the number of variates accepted compared to the total number generated, equals $1/k$. The rejection method may also be used for multivariate densities, except that the constant k is likely to have to be quite large, thus making this approach very inefficient. In practice it is difficult to ensure that condition 3.75 holds. Tierney [130] describes an approach, called a rejection sampling chain. This combines rejection sampling with Metropolis steps which means that the condition in equation 3.75 can be discarded.

The basic rejection algorithm can be improved upon greatly when the function $f(x)$ is expensive to compute. Rubenstein [108] introduces a second function $g(x) < f(x)$ that is inexpensive to compute. The candidate samples are first tested using $g(x)$, which if they pass are automatically accepted without having to evaluate $f(x)$. If the samples are rejected in the first test then they are given a second test using $f(x)$, which is the normal rejection algorithm. The first stage of this technique reduces the number of times $f(x)$ needs to be evaluated. This approach is known as the rejection method with squeeze. Gilks and Wild [39] have devised a method for adaptive rejection sampling. If $f(x)$ is a log-concave function then as more samples are drawn, the Gilks and Wild algorithm gives a piecewise linear squeeze function $g(x)$ that adaptively converges to $f(x)$ from below while the comparison function $c(x)$ converges to $f(x)$ from above.

3.7.5 Other methods

The method of composition [126] is a method for producing multidimensional random variates. Suppose it is desired to draw random variates (x_i, y_i, z_i) from a density $p(x, y, z) = c(x \mid y, z)\, b(y \mid z)\, a(z)$. The steps are:

1. $z_i \leftarrow a(z)$

2. $y_i \leftarrow b(y \mid z_i)$

3. $x_i \leftarrow c(x \mid y_i, z_i)$

This chained approach is fundamental to sampling algorithms such as the Gibbs sampler [36, 127] and data augmentation [126, 127].

Specialized techniques have been derived for simulating certain densities. The Box-Miller method [30, 95, 104] is an efficient method for generating Gaussian variates. Gamma variates with a certain number of degrees of freedom may be formed by summing together an appropriate number of exponential variates [95]. Cauchy variates may be generated as the ratio of zero mean unity variance Gaussian variates [30].

The most flexible approaches for producing random variates are based on Markov chain Monte Carlo methods. These include the Metropolis algorithm, the Gibbs sampler and the Hybrid Monte Carlo algorithm. These will be reviewed in some detail later because of their considerable importance in computational Bayesian methods.

3.8 Evidence using importance sampling

In this section, the aim is to devise an effective Monte Carlo method for the computation of Bayesian evidence.

As noted in section 3.6 the key to the successful application of importance sampling is to choose a sampling density that matches the integrand as closely as possible. In this example, the integrand is the product of likelihood times prior (times any Jacobian of transformation introduced by a change of coordinate system). Let the evidence be defined by the following integral:

$$p(\mathbf{d} \mid I) = \int_{\Re^M} f(\mathbf{x})\, d\mathbf{x} \tag{3.76}$$

where M is the dimensionality of the parameter space.

A little thought suggests that an approximately normal $g(\mathbf{x})$ would be an intelligent choice of sampling density to sample $f(\mathbf{x})$. Another element to the solution of the problem would be to transform to an orthonormal set of parameters. As described earlier, this transformation has the effect of making the integrand appear more "Gaussian" and of becoming an approximately separable function of the new parameter set.

3.8.1 *Choice of sampling density*

Let $\mathbf{e}_i$ be a random sample vector, each element of which is statistically independent. Let each element e_{ij} have a probability density $g_j(x)$. The probability density of the random vector $\mathbf{e}_i$ is therefore:

$$g(\mathbf{e}_i) = \prod_j g_j(e_{i\,j}) \tag{3.77}$$

From the point of view of numerical integration the optimum choice for $g_j(e_{i\,j})$ depends on the problem under consideration. Ideally, if the parameterization is perfect then all components of $f(x)$ will be Gaussian and therefore all $g_j(e_{i\,j})$ will be of the same form.

From the asymptotic normality principle, one may think that a Gaussian sampling density would be an appropriate choice for $g(x)$ in equation 3.68. This is not the case since the Gaussian density has rather thin tails. This leads to numerical instability [126] since the ratio $f(e_i)/g(e_i)$ in equation 3.68 can be extremely large in that part of the integrand about which least is known, namely the tails. A heavy tailed importance sampler avoids this problem.

Three reasonable choices for $g_j(x)$ are as follows:

Laplacian (two sided): $g_j(x) = \frac{1}{2}\exp[-|x|]$

The stages in generating Laplacian statistics are:

1.
$$u_1 \leftarrow (0,1] \tag{3.78}$$

2.
$$u_2 \leftarrow (0,1] \tag{3.79}$$

3.
$$x_i = \text{sign}(u_1 - 0.5)\log(u_2) \tag{3.80}$$

Hyperbolic Cauchy: $g_j(x) = \frac{1}{2\sqrt{2}}\,\text{sech}^2\left(\frac{x}{\sqrt{2}}\right)$

The stages in generating hyperbolic Cauchy statistics are:

1.
$$u_1 \leftarrow (0,1] \tag{3.81}$$

2.
$$x_i = \tanh^{-1}(2\,u_1 - 1) \tag{3.82}$$

Cauchy: $g_j(x) = \frac{1}{2\pi}\left[1 + (x/2)^2\right]^{-1}$

If the integrand is very heavy tailed then neither the Laplacian nor the hyperbolic Cauchy densities will be heavy tailed enough to give robust importance sampling. If this is the case then one could resort to using Cauchy (or Lorentzian) random variates instead:

1.
$$u_1 \leftarrow (0,1] \tag{3.83}$$

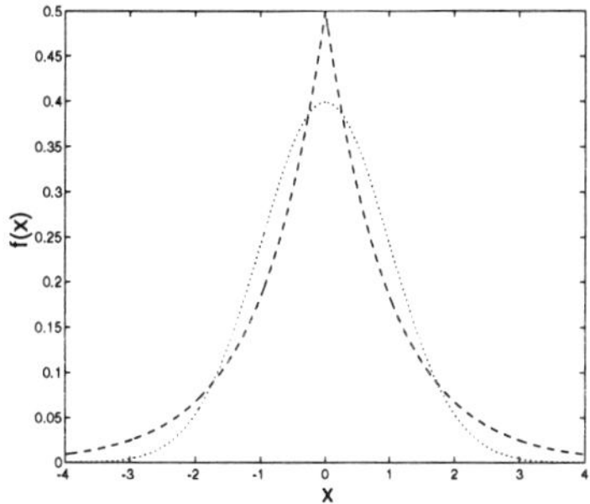

FIGURE 3.4. Laplacian density (dashed); Gaussian density (dotted)

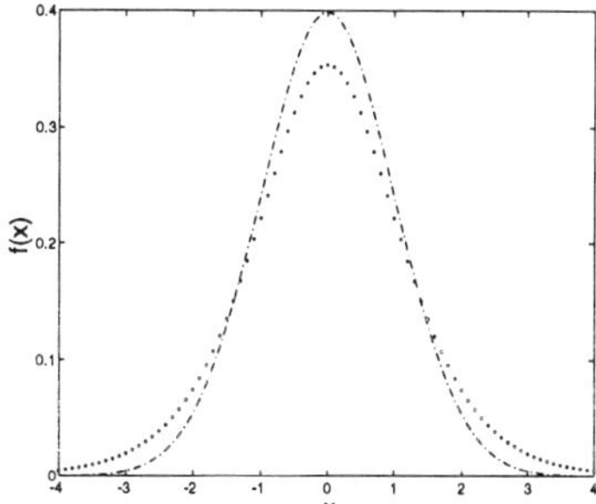

FIGURE 3.5. Hyperbolic Cauchy density (dot-dashed); Gaussian density (dotted)

2.

$$x_i = q = \sqrt{2}\,\tan\left[\pi\left(u - \frac{1}{2}\right)\right] \tag{3.84}$$

Alternatively, one may generate Cauchy variates as the ratio of two Gaussian variates as described by Devroye [30].

The above three transformations produce a random variate x_i whose density $g_j(x)$ is comparable to that of a unity variance Gaussian at the mode $x = 0$.

Figure 3.4 is the Laplacian density plotted against a unit variance Gaussian. The tails are obviously quite heavy but the resemblance between "integrand" and sampler is obviously very poor at the mode. Figure 3.5 is the hyperbolic Cauchy density plotted against a unit variance Gaussian. In this case we appear to have combined heaviness of tails with quasi-Gaussian behaviour in the region of the mode.

Smith [119] also considers choosing non-symmetric sampling densities $g(u)$ of the form:

$$g(u) = (1 - A)\,h(1 - u) + A\,h(u) \tag{3.85}$$

where u is a uniform variate and where $0 < A < 1$. The density $h(u)$ is of course non-symmetric. The degree of skewness in the density depends on the choice for the parameter A. Smith [119] mentions that A can be determined as part of the adaptive integration scheme described in section 3.4.1.

Lepage [72, 95] has developed the VEGAS algorithm for *adaptive importance sampling*. The basis of the method is to choose the importance sampling density to minimize the estimated variance of the Monte Carlo integral given by equation 3.69. The importance sampling density is assumed to be separable as above. Each component is updated iteratively according to a scheme described by Press *et al* [95] and Lepage [72]. This uses the expected variance of the integral as a cost function to choose the optimum importance sampler.

3.8.2 Orthogonalization using noise colouring

Now we discuss the incorporation of the mode and covariance information from the joint posterior density into the sample variates. We begin with uniform variates u_i which are then shaped using functional transformations to produce independent noise samples $w_i \in \Re$ with a density comparable to that of a zero mean unit variance Gaussian density. A vector $\mathbf{w}$ of random variates is required to sample the parameter space of dimensionality p. The probability density of this random vector equals the product of the probability densities of each (independent) component.

$$g(\mathbf{w}) = \prod_{i=1}^{p} g(w_i) \tag{3.86}$$

We have a vector $\mathbf{w} \in \mathbf{W}$ in transformed orthogonalized space. Let $\mathbf{x} \in \mathbf{X}$ be the corresponding vector in untransformed parameter space. Ideally we would like:

$$g\left(\mathbf{w}\right) \propto f\left(\mathbf{x}\right) \tag{3.87}$$

where $f(\mathbf{x})$ is the integrand.

The orthonormal transformation is defined as:

$$\mathbf{w}^{\mathbf{T}}\mathbf{w} = (\mathbf{x} - \hat{\mathbf{x}})^{\mathbf{T}}\mathbf{C}^{-1}(\mathbf{x} - \hat{\mathbf{x}}) \tag{3.88}$$

where $\mathbf{C}$ is the covariance matrix and $\hat{\mathbf{x}}$ is the mode of the normal approximation to the integrand.

The required variance and correlations between the parameters can be obtained by premultiplying the unit random vector by the square root of the covariance matrix. The square root may be obtained from the eigendecomposition of the covariance matrix. Let $\mathbf{P}$ be the matrix of eigenvectors and $\mathbf{\Lambda}$ be the associated diagonal matrix of (positive) eigenvalues. We may write:

$$\mathbf{C}^{\frac{1}{2}} = \mathbf{P}\,\mathbf{\Lambda}^{\frac{1}{2}} \tag{3.89}$$

This is exactly the same technique used to sample from a multivariate Gaussian density as shown in appendix B, except that in this case the vector $\mathbf{w}$ is not Gaussian distributed.

The joint posterior density may now be sampled about the mode $\hat{\mathbf{x}}$ of by adding the mode of the density to the coloured random vector.

$$\mathbf{x} = \mathbf{C}^{\frac{1}{2}}\,\mathbf{w} + \hat{\mathbf{x}} \tag{3.90}$$

The required integral may be estimated according to equation 3.68.

$$p(\mathbf{d} \mid I) = \int_{\Re^M} \frac{f(\mathbf{x})}{g(\mathbf{w})} g(\mathbf{w}) \left| \frac{\partial \mathbf{x}}{\partial \mathbf{w}} \right| d\mathbf{w} \approx \frac{1}{\sqrt{\det \mathbf{C}}} \frac{1}{N} \sum_{i=1}^{N} \frac{f(\mathbf{x}_i)}{g(\mathbf{w}_i)} \tag{3.91}$$

An important point to note is that the linear transformation introduces the Jacobian term as a constant scaling factor.

Overall, the effect is to sample about the mode of the joint posterior density, whilst both scaling and orientating the axes correctly, so that the importance sampling density resembles the integrand as closely as possible.

Equivalently, this sampling method may be understood in terms of a simple change of basis in the parameter space. The covariance matrix is symmetric which implies that it has a set of orthonormal eigenvectors. The square roots of the eigenvalues scale the axes of this orthogonalized space. Multidimensional white noise will sample this orthogonal space of scaled eigenvectors quite efficiently. This is equivalent to sampling the original non-orthogonal parameter space with coloured noise.

3.9 Marginal densities

Kloek and Van Dijk [65] have used adaptive importance sampling to compute marginal densities for parameter estimation.

In this section we shall present other importance sampling techniques to approximate marginal densities. There are two main kinds of approach:

- Produce random samples from the marginal density and form a histogram.
- Evaluate a continuous curve that is an estimate of the marginal density.

As we shall see, the first approach is obviously more primitive but it is easier to implement and more generally applicable. The second approach is obviously much more satisfactory, but it depends on being able to evaluate either the joint density or the conditional density as a continuous function.

The requirement that the sampling density be normalized may be ignored when approximating marginal densities. This gives complete freedom in the choice of sampling method. This is a key point in the discussion below.

3.9.1 Histograms

The most straightforward way to marginalize is to generate samples from the marginal density. These can be collected into frequency bins and the histogram plotted directly.

The usual procedure that is followed is to generate samples (x_i, y_i) from the *joint density* $f(x, y)$ and to ignore the variates y_i. Since the samples (x_i, y_i) are jointly distributed then it follows that the variates x_i are marginally distributed.

One could estimate the marginal density as a continuous function from its samples using either Parzen densities [128] or some other smoothing technique [116].

3.9.2 Jointly distributed variates

There are many approaches that may be used to generate jointly distributed variates. As mentioned before, Markov chain Monte Carlo methods such as the Metropolis algorithm, Gibbs sampler or Hybrid Monte Carlo may be used. These will be discussed in the next chapter.

At this stage we can also mention another method for simulating jointly distributed variates which is similar to the bootstrap approach [95]. The method of sampling importance resampling (SIR) of Rubin [109] is a non-iterative scheme for sampling from the joint density $f(x, y)$. The principle is as follows. Suppose that a population of N sample variates from a distribution $g(x, y)$ have been generated. By resampling from the population one can simulate drawing variates from $f(x, y)$ if each sample (x_i, y_i) is resampled from the population with weight w_i. The weights w_i are given by:

$$w_i = \frac{f(x_i, y_i)/g(x_i, y_i)}{\sum_i f(x_i, y_i)/g(x_i, y_i)} \tag{3.92}$$

The sampling-importance-resampling algorithm is really just importance sampling. Note that sample estimates produced using SIR are biased if the sample size is finite.

For our purposes, it is now exceedingly easy to generate approximate marginal densities from the population of sample variates x_i, y_i once evidence has been computed using importance sampling as detailed in section 3.8.

3.9.3 The dummy variable method

The dummy variable method is based on a principle similar to that of conditional Monte Carlo as described by Hammersley and Handscomb [47]. The essential idea is to introduce a dummy variable which increases the dimensionality of the integral. This should make the problem of integrating to obtain the marginal density the same as integrating to obtain the evidence. In this way we can convert the new problem of evaluating a marginal density into a problem that has been solved already. Chen [23] uses what essentially amounts to the same trick for computing marginal densities.

Let the parameter space consist of two subspaces $\mathbf{X}$ and $\mathbf{Y}$. We are interested in computing marginal densities of the form:

$$p(x) = \int_{\mathbf{Y}} p(x, y)\, dy \tag{3.93}$$

where $y \in \mathbf{Y}$ is $p \times 1$ and $x \in \mathbf{X}$ is $q \times 1$. Suppose also that a function $h(x, y)$ is defined such that:

$$H(y) = \int_{\mathbf{X}} h(x, y)\, dx \tag{3.94}$$

We can therefore write equation 3.93 in the form:

$$p(x) = \int_{\mathbf{Y}} \frac{\int_{\mathbf{X}} h(u, y)\, du}{H(y)}\, p(x, y)\, dy \tag{3.95}$$

Using Fubini's theorem this simplifies to:

$$p(x) = \int_{\mathbf{X}} \int_{\mathbf{Y}} \frac{p(x, y)\, h(u, y)}{H(y)}\, dy\, du \tag{3.96}$$

Note that we have increased the dimensionality of the integral from p in equation 3.93 to $p + q$ in equation 3.96. In general, integrals become *more difficult* as the dimensionality is increased. It would therefore appear that it is not a very good idea to increase the dimensionality in the manner described. However, the above can lead to some considerable simplifications as will be shown below.

The choice for $h(u, y)$ is actually completely arbitrary; all that is required is that it is non-zero for all points where $p(x, y)$ is non-zero and that equation 3.94 holds. We have already computed the evidence. Therefore let us choose $h(u, y)$ so that the joint density is *approximately proportional* to the integrand of equation 3.96:

$$p(u, y) \tilde{\propto} \frac{p(x, y)\, h(u, y)}{H(y)} \tag{3.97}$$

A good choice for $h(u, y)$ is the normal approximation to the integrand $p(u, y)$. The corresponding marginal density $H(y)$ may be found using the results derived in section 3.1.6.

It follows that this expression may be integrated using the same importance sampling density as was used previously for evaluating the evidence in section 3.8. Let us denote the importance sampling density used to compute the evidence as $g(u, y)$.

$$p(x) = \int_X \int_Y \frac{p(x, y)\, h(u, y)}{H(y)\, g(u, y)}\, g(u, y) dy\, du \tag{3.98}$$

$$\approx \frac{1}{N} \sum_{i=1}^{N} \frac{p(x, y_i)\, h(u_i, y_i)}{H(y_i)\, g(u_i, y_i)} = \frac{1}{N} \sum_{i=1}^{N} \frac{p(x, y_i)\, h(u_i \mid y_i)}{g(u_i, y_i)} \tag{3.99}$$

where the sample points u_i, y_i are distributed with density $g(x, y)$. Note that the right hand side is a continuous function of the parameter x. All the sample variates u_i, y_i are already generated so it is simply a matter of inserting x into equation 3.98 in order to estimate the marginal density.

This technique is very interesting for a number of reasons. First, it assumes that evidence has already been evaluated, and as such a large number of function evaluations have already been carried out. The aim is to recycle as much of the information obtained in calculating the evidence as possible. Second, the true direction of Bayesian inference should be to select the model first followed by computing the marginal density later. It makes little sense computing marginal densities until the model has performed well against other competing models. This technique is consistent with the Bayesian way of thinking.

3.9.4 *Marginalization using jointly distributed variates*

An interesting special case of the dummy variable method arises if we make the choice:

$$g(x, y) = h(x, y) = p(x, y) \tag{3.100}$$

When this is the case the integral becomes:

$$p(x) = \int_{\mathbf{Y}} \int_{\mathbf{X}} p(x \mid y)\, p(u, y)\, du\, dy \tag{3.101}$$

$$\approx \frac{1}{N} \sum_{i=1}^{N} p(x \mid y_i) \tag{3.102}$$

where the sample points (u_i, y_i) are jointly distributed with density $p(x, y)$. The use of formula 3.101 depends on two things. First, it must be possible to draw random variates (u_i, y_i) from the joint density. Second, the conditional density must be available as a function of the parameters. Note that equation 3.102 is a continuous function of the parameter x providing the conditional density is itself a continuous function of the parameter x. Appendix E describes a numerical approach to computing expression 3.102 given the joint density in functional form.

3.10 Opportunities for variance reduction

Hammersley and Handscomb [47] discuss a number of different variance reduction techniques (or "swindles") which may be used in parallel with importance sampling. These include the use of non-random sequences and antithetic variates.

3.10.1 *Quasi-random sequences*

Press *et al* [95] note that it is not particularly efficient to use purely random variates for numerical integration because the standard deviation of the estimated integral is of order $\mathcal{O}(N^{-1/2})$. This rate of convergence is quite poor, which means there is very little point in carrying out very many function evaluations to do an integral. The essential problem they identify is that random numbers tend to cluster. Therefore for a large number of function evaluations carried out over a given range, many of the function evaluations will take place at positions that are too close to previous evaluations to contribute much information about the value of the integral.

Hammersley and Handscomb [47] and Press *et al* [95] propose using quasi-random sequences instead. These sequence are cleverly constructed to be *self avoiding* which means new variates are positioned as far away as possible from previous variates. The estimated standard deviation is of order $\mathcal{O}(1/N)$. The simplest quasi-random sequences are equidistributed [47] which means one should use regularly spaced samples with uniform random phase. Number theory may be used to produce sequences with the self avoiding property. Halton sequences [47] are the simplest example. Sobol' sequences [95] are more sophisticated and have better lattice structure in the sense that Sobol' sequences fill the space more efficiently. The implementation of Sobol' sequences is trivial, since all that is required is for the uniform number generator (used to draw samples in parameter space for function evaluations) to be replaced with a Sobol' sequence generator.

Quasi-random numerical integration for Bayesian applications is discussed by Shaw [115].

3.10.2 *Antithetic variates*

The antithetic variates technique involves constructing sequences of random samples, in such a way as to introduce negative correlations. Recall that all random simulations, including Monte Carlo integration, need a source of uniform random numbers. These are shaped and transformed and the integrand evaluated. For one sequence consisting of random variates u_i one could produce a sequence of unbiased estimates $\theta_k = f(u_i)/g_k(u_i)$ of the integral. One example of the technique known as antithetic variates is then to compute a second sequence of unbiased estimates of the integral $\theta_k = f(e_i)/g_k(e_i)$ using a sequence of uniform variates $e_i = 1 - u_i$. The integral is computed as the average of the two (negatively correlated) sequences.

Hammersley and Handscomb [47] show that the variance of the sum of two antithetic sequences is always less than the expected sum if the sequences were independent (i.e. the sum of the variances). Geweke [38] discusses an example of the application of this technique to Bayesian integrals. In this example, the integrand is almost symmetric. Antithetic variates are constructed such that the sampling is symmetric about the mode of the distribution. Intuitively the effects of overestimating the integral on one side is compensated by underestimating the value on the other side. Hammersley and Handscomb [47] extend the principle and discuss some rather more sophisticated techniques that use correlations between sequences of unbiased estimates of the integral for variance reduction.

3.10.3 Control variates

If one considers the use of antithetic variates as variance reduction using negative correlation, then one can consider the use of control variates as variance reduction by positive correlation. A set of random variates is used to evaluate two integrals, one integral $\hat{I}$ whose value is unknown and another integral $\hat{C}$ whose value is known exactly. Both integrals can be estimated using Monte Carlo methods. The error in the estimate for the second integral is exactly known.

Rubenstein [108] proposes a linear relationship of the form:

$$I - \hat{I} = \beta_1(C_1 - \hat{C}) \tag{3.103}$$

The coefficient β_1 is chosen in order to compensate for the errors in estimating the integral due to the random sample realization used in evaluating the integrand. The idea may be extended to more than one control integral, each corresponding to a coefficient β_i. Control integrals may be the moments of the sampling density, or any other function. For the example in equation 3.103 a good choice for C is an approximation (possibly Gaussian) to I.

The coefficients β_i can be determined by *training* over a large number of trial integrals. This approach is not always very good in practice, because the corrections can only be as good as the best estimate for I obtained by using the training set. The method does not seem to work well for multivariable problems because it is difficult for a linear scheme such as equation 3.103 to compensate for all the variations in all the variables at once. Despite its shortcomings for actually estimating integrals, the method of control variates does find some use in assessing the convergence of a Markov chain, as described by Tanner [126].

3.10.4 Stratified sampling

The method of stratified sampling involves splitting the region of integration into separate regions and integrating over each region separately. Hammersley and Handscomb [47] show that the optimum number of function evaluations needed to integrate over each region depends on the variability of the integrand in that region. Adaptive stratified sampling [95] is based on this principle. We do not consider this a worthwhile approach to Bayesian integration because the number of regions increases exponentially with the number of dimensions.

3.11 Summary

In this chapter, various approaches to numerical Bayesian inference were described. Solutions to three areas in Bayesian inference were reviewed; namely given a probability density, to locate the supremum, to draw random samples from it and to compute marginal densities and evidence.

The posterior density is assumed to consist of a single dominant mode around which most of the probability mass is concentrated. In addition to the dominant mode there may also be many local modes. The presence of these local modes means that standard gradient based approaches to optimization may be completely ineffective for many problems. Global algorithms such as simulated annealing or genetic algorithms are required in these cases. If the current position actually lies within the region occupied by the dominant mode, then a local quadratic algorithm such as conjugate gradients or variable metric should work because of the highly quadratic nature of the log density in the region of the mode. One may decide to combine both global and local algorithms into a hybrid strategy; using a global algorithm initially to locate the dominant mode and then switching to a local algorithm once inside the region of the mode.

The fundamental problem of computing Bayesian integrals, such as marginal densities, evidence and posterior expectations was addressed. The approaches considered included standard numerical quadrature, various asymptotic approximations and Monte Carlo integration. Numerical quadrature was found to be ineffective for computing integrals of high dimensionality. Asymptotic approximations were proposed that exploit the quasi-Gaussian nature of Bayesian integrands. These require numerical optimization for conditional maximization of the density, and numerical differentiation to determine the curvature of the density about the conditional modes. The accuracy of asymptotic approximations depends on the number of data and is always subject to systematic error.

The main approach to the computation of Bayesian integrals is Monte Carlo integration. Monte Carlo integration has many features that distinguish it from other methods of integration. First, the integral is *estimated* as the mean (or some other statistic) of a random process. This estimate is (usually) unbiased and repeated trials will lead to increasing accuracy in the estimate. The convergence of the estimate is rather slow, being of the order of $\mathcal{O}(N^{1/2})$. Second, the method does not depend directly on the dimensionality, which means that it can be used to reliably estimate large scale multidimensional integrals.

Several Monte Carlo techniques were devised for integrating the posterior density to obtain marginal densities and evidence. These include the dummy variable method for computing marginal densities and an importance sampling technique for computing evidence. Monte Carlo integration requires methods for simulating from prescribed densities. Therefore, common approaches to sampling from probability densities were discussed. The chapter concluded with a review of variance reduction techniques that can be used in conjunction with sampling methods to increase the accuracy of sample estimates. The most effective means of variance reduction is to use a good importance sampling density. Optimally, one should sample from the integrand, which in our case is the joint density. Markov chain Monte Carlo techniques will provide a means of carrying this out.

4

Markov Chain Monte Carlo Methods

4.1 Introduction

Sampling methods based on Markov chains were first developed for applications in statistical physics. Two branches of development originated in the 1950s. The classic paper by Metropolis *et al* [77] introduced what is now known as the *Metropolis algorithm*. This method was popularized for Bayesian applications, along with its variant the Gibbs sampler, by the influential papers of Geman and Geman [36], who applied it to image processing, and Gelfand and Smith [35], who demonstrated its application to Bayesian problems in general.

At the same time as the Metropolis algorithm was introduced, Alder and Wainwright [3] developed the "molecular dynamics" method, in which new states are found by simulating the molecular dynamics of a system. This method can be used to simulate from any differentiable distribution and is called the dynamical method. More recently, the two branches have converged to produce the *Hybrid Monte Carlo* [4, 82] method.

There are two key concepts in this chapter which underlie the main body of the work from this field.

Theory of Markov chains: A Markov chain is the first generalization of an independent process. Each state of a Markov chain depends on the previous state only. This idea may be applied to modelling the behaviour of molecules in a gas. The energy of a molecule depends only on the last collision. This model is used to simulate the motion of a hypothetical "molecule" through parameter space.

Energy and probability: The canonical distribution of the energy of a molecule is used exclusively in this work. Relating energy to probability in this way leads to powerful dynamical methods for exploring parameter space.

4.2 Background on Markov chains

The aim in this section is to review some of the key concepts behind Markov chain Monte Carlo methods with a view to constructing Markov chains for sampling specific densities, and hence compute marginal densities, posterior expectations and evidence.

As noted earlier, the most flexible approaches to the generation of random variates with a prescribed density are based on Markov chains. The theory of Markov chains is well developed and is presented by many authors such as Kalos and Whitlock [59], Kennedy [61], Tierney [130] and Neal [82].

A Markov chain is defined [82] to be a series of random variables x_1, x_2, x_3, ... x_N such that their influence on the value of x_{N+1} is mediated by the value of x_N alone:

$$p\left(x_{N+1} \mid x_1, x_2, x_3, \ldots x_N\right) = p\left(x_{N+1} \mid x_N\right) \tag{4.1}$$

The random variables have a common range known as the *state space* of the Markov chain. State space variables may be either discrete or continuous. The set of discrete space variables may be finite or countably infinite. In most cases[1] the set of state space variables is identical to the parameter space.

A Markov chain begins with an initial distribution for x_0, and thereafter the distribution of x_i is determined by the transition probabilities between states:

$$p_{N+1}(x) = \sum_{\{x'\}} p_N(x')\, T_N(x', x) \tag{4.2}$$

We denote the initial distribution as $p_0(x)$. In continuous state spaces the summation is replaced by an integral.

The function T_N is known as a *base transition*. In some Markov chains the base transition does not depend on N and the Markov chain is said to be homogeneous (e.g. the global Metropolis algorithm). In other examples that we encounter, each base transition only affects one component of the state space vector $\mathbf{x}$ at a time and the base transitions are rotated in sequence (e.g. Local Metropolis algorithm, Gibbs sampler).

---

[1]The Hybrid Monte Carlo state space is a superset of the model parameter space.

There are two properties required of the Markov chain for it to be of any use in sampling a prescribed density. First, there must exist a unique invariant distribution. Second, the Markov chain must be ergodic.

Invariant distribution: An invariant distribution is a fixed point solution to equation 4.2. In other words, the invariant distribution $\pi(x)$ is invariant with respect to the Markov chain if it solves:

$$\pi(x) = \sum_{\{x'\}} \pi(x') T_N(x', x) \tag{4.3}$$

for all x', x in parameter space.

In particular we are interested in *time reversible* Markov chains which satisfy the more restrictive condition of *detailed balance*:

$$\pi(x) T_N(x, x') = \pi(x') T_N(x', x) \tag{4.4}$$

for all pairs (x, x') in state space. Stated in words, the probability of a transition from state x to state x' is equal to the probability of a transition in the reverse direction. It is easy to show that detailed balance implies invariance.

Ergodicity: We also require that the Markov chain be *ergodic*. A Markov chain is said to be ergodic [82] if, regardless of the initial distribution, the probabilities at time N converge to the invariant distribution as $N \to \infty$.

$$\lim_{N\to\infty} p_N(x) = \pi(x) \tag{4.5}$$

If a Markov chain is ergodic then its invariant distribution is said to be an *equilibrium distribution.*

The rate of convergence of a Markov chain, or indeed whether it converges at all, is of crucial interest. How long it takes before the difference is negligible between the current distribution $p_N(x)$ and the equilibrium distribution depends on a number of factors. This is related to the number of states that must be discarded at the initial stage, a transient period known as the "burn in" of the Markov chain.

In theory, the analysis of convergence is trivial when the state space is finite because the transition probabilities from state x_N to x_{N+1} can be written in matrix form:

$$\mathbf{x}_{N+1} = \mathbf{T}\,\mathbf{x}_N \tag{4.6}$$

It can be shown [114] that the rate of convergence depends on the magnitude of the second largest[2] eigenvalue of $\mathbf{T}$. A value just less than unity gives

[2]The largest eigenvalue is always equal to one. The equilibrium distribution is given by the eigenvector corresponding to the eigenvalue with value one.

relatively slow convergence of the Markov chain. In practice, of course, it is usually not feasible to determine the eigenvectors and eigenvalues because of the high dimensionality of state space.

The analysis of convergence of Markov chains in continuous parameter spaces is much more difficult. A theoretical treatment is given by Nummelin [84]. Kalos and Whitlock [59] present an analysis of a simple example.

From our point of view, we are usually interested in the simulation of prescribed densities on a digital computer. Digital computers do not actually use real numbers but represent them to a fixed precision in floating point form. Therefore we shall be satisfied if we can show that the prescribed Markov chain converges for a discrete state space.

Neal [82] presents a remarkably simple proof of the following theorem.

Theorem 1: *Let a Markov chain have the invariant distribution* $\pi(\mathbf{x})$ *and define* ν *such that:*

$$\nu = \min_{\{\mathbf{x}\}} \min_{\{\mathbf{x}':\pi(\mathbf{x},\mathbf{x}')>0\}} \frac{T(\mathbf{x},\mathbf{x}')}{\pi(\mathbf{x})} > 0 \tag{4.7}$$

where $T(\mathbf{x},\mathbf{x}')$ *is the transition probability of the Markov chain.*

1. *If the probability distribution at time* n *is given by a mixture density*

$$p_n(\mathbf{x}) = \left(\mathbf{1} - (\mathbf{1}-\nu)^{\mathbf{n}}\right) \pi(\mathbf{x}) + (\mathbf{1}-\nu)^{\mathbf{n}} \mathbf{r_n}(\mathbf{x}) \tag{4.8}$$

then the probability distribution at time $n+1$ *is of the form:*

$$p_{n+1}(\mathbf{x}) = \left(\mathbf{1} - (\mathbf{1}-\nu)^{\mathbf{n+1}}\right) \pi(\mathbf{x}) + (\mathbf{1}-\nu)^{\mathbf{n+1}} \mathbf{r_{n+1}}(\mathbf{x}) \tag{4.9}$$

where $r_n(\mathbf{x})$ *and* $r_{n+1}(\mathbf{x})$ *are normalized probability densities.*

2. *Bounds on the rate of convergence:*

$$|\pi(\mathbf{x}) - \mathbf{p_n}(\mathbf{x})| < (1-\nu)^n \tag{4.10}$$

3. *Convergence of expected values:*

$$|\langle a \rangle - \mathrm{E}_N[a]| \leq (1-\nu)^n \max|a(\mathbf{x}) - a(\mathbf{x}')| \tag{4.11}$$

Equation 4.9 is the most important from our point of view. The right hand side of equation 4.9 consists of two components. As n increases the weight on the invariant distribution approaches one, while the weight on the impurity $r_{n+1}(\mathbf{x})$ will approach zero.

This means that providing the Markov chain is *regular*, that is $T(\mathbf{x},\mathbf{x}') > 0 \;\; \forall \;\; (\mathbf{x},\mathbf{x}')$ and homogeneous, and providing an invariant distribution can be shown to exist, the Markov chain will be ergodic. In the experimental section of this book we shall only consider Markov chains that are regular.

We prove in appendix F that the three main sampling techniques; namely, the Metropolis algorithm, Gibbs sampler and Hybrid Monte Carlo all satisfy detailed balance. This implies that an invariant distribution exists in each case and is identical to the joint density.

Theorem 1 as stated applies only to homogeneous Markov chains. This is a potential problem since, for example, the Gibbs sampler is not a homogeneous Markov chain. In fact it is *cyclic* which means that the base transitions repeat periodically. As noted by Neal [82], Theorem 1 still applies since each period in the cycle (or iteration) may be regarded as a single base transition of a homogeneous Markov chain.

4.3 The canonical distribution

The canonical distribution was originally devised in statistical mechanics to describe the effects of temperature on the large scale behaviour of the molecules of a gas. The key quantities that come into play are energy and temperature. How the probability density of states depends on these two quantities is of immense interest and is the foundation of optimization using simulated annealing, sampling using dynamical methods as well as the computation of evidence using free energy methods.

4.3.1 Energy, temperature and probability

Molecules in a gas move in random directions in space with an average kinetic energy proportional to the average temperature. Each molecule moves in a straight line until it encounters another molecule and a collision occurs. The random zigzag path that results from such intermolecular collisions is visible when pollen grains are examined under a microscope, and is known as Brownian motion, after its discoverer the Scottish botanist Robert Browne.

The equilibrium probability distribution for the energy of the ensemble of molecules is known as the canonical distribution and is given by

$$P(E) = \frac{1}{Z} \exp\left[-\frac{E}{kT}\right] \tag{4.12}$$

where T is the absolute temperature, k is Boltzmann's constant and the normalization constant Z is known as the partition function. From the above one can see that the expected energy of each molecule equals $\langle E \rangle = kT$ and that the occupancy of higher energy level states increases with higher temperature. States that are virtually inaccessible at low temperature can become very probable at higher temperatures. In all cases the state of highest probability is the state of lowest energy $E = 0$. In fact, for the special case $T = 0$ this is the only possible state.

For our purposes, we may choose any definition for the energy. Since, in nature, energy is generally minimized then it obviously makes sense to equate energy to something that we wish to minimize. We wish to maximize the posterior probability $p(\{\omega\} \mid \mathbf{d}, I)$ where $\{\omega\}$ is the set of model parameters, so we will find it useful to define energy as minus the natural logarithm of the posterior probability. Since the temperature scale is arbitrary then for convenience we will set $k = 1$ which means that the canonical distribution becomes:

$$P_B(E(\{\omega\})) = \frac{1}{Z} \exp\left[\frac{1}{T} \log\left(p(\{\omega\} \mid \mathbf{d}, I)\right)\right] \tag{4.13}$$

where Z is the partition function of the system. The energy function is a function of the model parameters, which constitute the states of the system. The parameter space is equivalent to the set of possible states that can be attained by this hypothetical thermodynamic system.

The posterior distribution is a special case of the canonical distribution. Setting $T = 1$ in equation 4.13 yields the desired result. With $T = 1$ it is also clear that the partition function is exactly equivalent to model evidence.

$$Z|_{T=1} = \int_\Omega \exp\left[\log\left(p(\{\omega\} \mid \mathbf{d}, I)\right)\right] d\omega = \int_\Omega p(\{\omega\} \mid \mathbf{d}, I)\, d\omega \tag{4.14}$$

Later, it will be shown how to determine model evidence using statistical mechanical methods for determining free energy.

4.3.2 Random walks

The sequence of random variables from a Markov chain tend to move through state space in such a way that each sample from the sequence is highly correlated with the previous sample. This is the picture that one obtains of a particle or molecule being buffeted by surrounding particles and moving through space in a jerky zigzag pattern. The term "random walk" or "drunk man's walk" is often used to describe such behaviour. Neal [82] analyses a simple example of such a sequence:

$$\begin{aligned} T(k, k-1) &= \frac{1}{4} \\ T(k, k) &= \frac{1}{2} \\ T(k, k+1) &= \frac{1}{4} \end{aligned}$$

It is easy to show that the variance of the sequence is $N/2$ where N is the number of steps (i.e. time). The expected distance that this sequence will diffuse in fixed time is therefore proportional to the square root of time.

In other words, the time expected to move over a certain distance varies as the square of the distance. Generalizing to other examples of random walks, Neal concludes that random walks are quite inefficient at *exploring* parameter space, and that if the starting point is located at too great a distance from the probability mass then the Markov chain is unlikely to reach equilibrium in a reasonable amount of time.

4.3.3 Free energy and model selection

The free energy F is defined as

$$F = -\log Z \tag{4.15}$$

where Z is the partition function. It is trivial to show that free energy is related to the energy of the state by

$$F = \langle E \rangle - T\,S \tag{4.16}$$

where $\langle E \rangle$ is the expected energy and where S is known as the entropy of the system:

$$S = -\int_{\mathbf{X}} p(\mathbf{x}) \log p(\mathbf{x})\, d\mathbf{x} \tag{4.17}$$

Free energy is commonly used in physical chemistry as a means of determining the tendency of a chemical reaction to occur. A given chemical reaction will occur in preference to another possible reaction if the free energy is lower.

This corresponds rather nicely with Bayesian inference. We defined free energy as minus log evidence. In Bayesian model selection one chooses from among the possibilities according to increasing model evidence, or equivalently, decreasing free energy.

4.4 The Gibbs sampler

The Gibbs sampler is perhaps one of the most flexible sampling techniques available. The Gibbs sampler became prominent as a result of the work of Geman and Geman [36] in image processing, and Gelfand and Smith [35] in data analysis.

The principle underlying the use of the Gibbs sampler is that one can break down the problem of drawing samples from a multivariate density into one of drawing successive samples from densities of smaller dimensionality. In its usual form the Gibbs sampler draws samples from univariate densities.

4.4.1 Description

Assume for example that the parameter space consists of k components $\{a_1, a_2, a_3 \ldots a_k\}$. The components are initialized to starting values $\{a_1^0, a_2^0, a_3^0 \ldots a_k^0\}$.

The Gibbs sampler proceeds by drawing random variates from conditional densities in a cyclical iterative pattern as follows:

First iteration:

$$\begin{array}{lcl} a_1^1 & \leftarrow & p\,(a_1 \mid a_2^0\, a_3^0 \ldots a_{k-1}^0\, a_k^0) \\ a_2^1 & \leftarrow & p\,(a_2 \mid a_3^0\, a_4^0 \ldots a_k^0\, a_1^1) \\ a_3^1 & \leftarrow & p\,(a_3 \mid a_4^0\, a_5^0 \ldots a_1^1\, a_2^1) \\ \vdots & \vdots & \vdots \\ a_k^1 & \leftarrow & p\,(a_k \mid a_1^1\, a_2^1 \ldots a_{k-2}^1\, a_{k-1}^1) \end{array}$$

Second iteration:

$$\begin{array}{lcl} a_1^2 & \leftarrow & p\,(a_1 \mid a_2^1\, a_3^1 \ldots a_{k-1}^1\, a_k^1) \\ a_2^2 & \leftarrow & p\,(a_2 \mid a_3^1\, a_4^1 \ldots a_k^1\, a_1^2) \\ a_3^2 & \leftarrow & p\,(a_3 \mid a_4^1\, a_5^1 \ldots a_1^2\, a_2^2) \\ \vdots & \vdots & \vdots \\ a_k^2 & \leftarrow & p\,(a_k \mid a_1^2\, a_2^2 \ldots a_{k-2}^2\, a_{k-1}^2) \\ \vdots & \vdots & \vdots \end{array}$$

n^{th} iteration:

$$\begin{array}{lcl} a_1^n & \leftarrow & p\,(a_1 \mid a_2^{n-1}\, a_3^{n-1} \ldots a_{k-1}^{n-1}\, a_k^{n-1}) \\ \vdots & \vdots & \vdots \end{array}$$

where $p\,(a_i \mid a_j\; j \neq i)$ denotes the conditional density of the ith component of the vector $\mathbf{a}$. The superscript number denotes the current iteration. Note that as soon as a variate is drawn, then it is inserted immediately into the conditional probability density function, and it remains there until it is substituted in the next iteration.

At the end of the jth iteration the sample $\{a_1^j, a_2^j, a_3^j \ldots a_{k-1}^j, a_k^j\}$ is considered to be a sample from the joint density. In common with other Markov chain approaches the Gibbs sampler requires an initial transient period to converge to equilibrium. How much of the initial series is affected by the initial state is difficult to ascertain, but some literature is available on the subject [104]. This initial period of length M is known as the "burn in" and it varies in length depending on the problem. One should always discard the first M samples as garbage.

Metropolis subchains

If the conditional densities are easy to sample from then the Gibbs sampler will be easy to implement. In many cases some of the densities are not of a simple standard form. In such cases one can resort to using the Metropolis algorithm as a means of drawing samples from the density. The idea of using Metropolis subchains was suggested by Müller (see Tanner [126]) and by Zeger and Karim [141]. This form of Gibbs sampler may be regarded as a hybrid strategy between the true Gibbs sampler and the local Metropolis algorithm, where "Gibbs steps" which exploit easily sampled conditional densities are interleaved between "Metropolis steps" which sample more difficult conditional densities.

4.4.2 Discussion

Tierney [130] refers to the Gibbs sampler as "sampling by conditioning". The Gibbs sampler is to the sampling of multivariate densities as the alternating variables method is to the optimization of multivariable cost functions. Not surprisingly, the Gibbs sampler suffers from the same disadvantages as the alternating variables method, namely that its rate of convergence is mainly governed by *a posteriori correlations* between the different parameters and the *dimensionality* of the parameter space [119]. The rate of convergence can be improved by means of some appropriate reparameterisation, as discussed by Hills and Smith [50].

A pdf whose contours are similar to those figure 3.2 presents difficulty because successive variates are very highly correlated. Smith [119] illustrates this by considering the effects of the correlation coefficient when using the Gibbs sampler to sample a bivariate Gaussian distribution. Smith describes how this particular weakness in the Gibbs sampler can be greatly alleviated by applying orthogonalizing transformations to the joint density.

Large dimensionality is also a problem because many base transitions are required to execute one iteration of the Gibbs sampler and hence produce a single sample from the joint density. In addition, the more parameters there are in a system the greater the probability that any two will be highly correlated[3].

4.4.3 Convergence

In appendix F.1, it is easily shown that the Gibbs sampler satisfies detailed balance and that the joint density is thus an invariant distribution of the Markov chain.

From Theorem 1 given in section 4.2, providing the parameter space is discrete, it is sufficient for the joint density to be positive everywhere in

[3]Intuitively, one can expect this to be especially true if the data are overfitted.

parameter space (i.e. the Markov chain is regular) for the Markov chain to be ergodic. More general proofs on convergence of the Gibbs sampler and uniqueness of the invariant distribution are given by Geman and Geman [36].

4.5 The Metropolis-Hastings algorithm

The Metropolis-Hastings algorithm, which is described by Kalos and Whitlock [59], is an extremely flexible method for producing a random sequence of samples from a given density. The Metropolis algorithm [4] was originally introduced by Metropolis *et al* [77] for computing the properties of substances composed of interacting molecules. This algorithm has been used extensively in statistical physics [47]; its use being so widespread that in some circles the term has become almost synonomous with Monte Carlo work. A generalization introduced by Hastings [48] and extended by Peskun [91] is presented below. The standard form of simulated annealing algorithm [63, 64, 95] uses the Metropolis algorithm, with the result that a large proportion of the optimization community do not distinguish between simulated annealing and the Metropolis algorithm. The Gibbs sampler can be viewed as a special case of the Metropolis-Hastings algorithm. An excellent review is presented by Neal [82].

4.5.1 The general algorithm

In this section we describe the Metropolis-Hastings algorithm in its more general form as described by Hastings [48]. It is assumed that the density is given in a functional form $y = p(x)$, where $x \in \Phi$. The density does not need to be normalized. The algorithm explores the parameter space Φ by means of a random walk. Suppose X_i is the ith element of such a random walk. It is proposed that the next variate in the random sequence be Y_i which is produced by adding a random perturbation ζ to X_i:

$$Y_i = X_i + \zeta \tag{4.18}$$

where ζ is drawn from the proposal density $s(\zeta)$.

There are two possibilities for the choice of the next variate in the random sequence:

- $X_{i+1} = Y_i$ **Accept** the proposed random variate.
- $X_{i+1} = X_i$ **Reject** the proposal and repeat X_i.

---

[4]The Metropolis algorithm is a relatively simple special case of the so called "Metrolis-Hastings" algorithm.

The probability of accepting Y_i instead of the current value is given by the Metropolis-Hastings[5] acceptance function:

$$A(X_i, Y_i) = \min(1, Q(X_i, Y_i)) \tag{4.19}$$

where

$$Q(X_i, Y_i) = \frac{p(Y_i)\, T(Y_i \mid X_i)}{p(X_i)\, T(X_i \mid Y_i)} \tag{4.20}$$

The conditional probability density $T(Y_i \mid X_i)$ is identical to the proposal density $s(\zeta)$. Therefore $T(X_i \mid Y_i)$ is given by the probability density $s(-\zeta)$. Let ϵ be a uniform random variate drawn over the range [0, 1]. If the condition[6]:

$$p(Y_i)\, T(Y_i \mid X_i) > \epsilon\, p(X_i)\, T(X_i \mid Y_i) \tag{4.21}$$

holds then the next term X_{i+1} in the random sequence is $X_{i+1} = Y_i$, otherwise we have $X_{i+1} = X_i$.

Clearly, if the proposal density $s(\zeta)$ is symmetrical about the origin then the acceptance probability 4.20 above is given by

$$Q(X_i, Y_i) = \frac{p(Y_i)}{p(X_i)} \tag{4.22}$$

Note that the probability of acceptance is independent of the probability density $s(\zeta)$. Also we can see that a position of lower probability will be always be exchanged for a position of higher probability. However, a transition in the reverse direction depends on a favourable outcome to a random event simulated using a uniform random deviate. Intuitively we can see that this will simulate the density $p(x)$ because, at equilibrium, the density of occupation of the states will be in proportion to the probability density. Before equilibrium is attained the system will tend towards states of higher probability. This is the original form of the Metropolis algorithm [77].

In fact, we shall distinguish between two forms of the Metropolis algorithm. In the case of the global Metropolis algorithm all components of the vector $\mathbf{X}$ will be updated in expression 4.18 above. In the local Metropolis algorithm, only one component will be updated at a time, and the base transitions (or "jumps") will be rotated in sequence in the same manner as in the Gibbs Sampler in section 4.4 above.

---

[5] Alternatively, one can use the Boltzmann acceptance function [82] but this is less efficient.

[6] A small computational saving can be made if one recognizes that $\epsilon \leq 1$. In other words, if $p(Y_i)\, T(Y_i \mid X_i) > p(X_i)\, T(X_i \mid Y_i)$ then Y_i is automatically accepted as the next variate in the random sequence, so the random variate ϵ need not be generated.

4.5.2 Convergence

In appendix F.2 it is shown that the Metropolis-Hastings algorithm satisfies detailed balance.

As in the case of the Gibbs sampler in section 4.4 we shall be satisfied with showing the Metropolis algorithm is ergodic in discrete parameter space. Since detailed balance holds, then the only condition required is that the proposal density is strictly positive everywhere. This will be the case for all proposal densities used in this work.

A full analysis of the behaviour of the Metropolis algorithm in continuous parameter spaces, as stated earlier, is difficult. Kalos and Whitlock [59] present a simple example where a one dimensional continuous Markov chain produced by the Metropolis algorithm using a uniform proposal density converges to the desired equilibrium distribution $y = 2x \quad x \in [0, 1]$. The rate of convergence is geometric, which is also the case for discrete Markov chains. A rigorous theoretical treatment of the convergence of the Metropolis algorithm in continuous spaces is presented by Nummelin [84].

4.5.3 Choosing the proposal density

The random sequence produced by the Metropolis algorithm is in the form of a random walk in which successive sample variates are highly correlated. We have shown that the Metropolis-Hastings algorithm samples the required distribution under general conditions. However, we have not addressed the issue of ensuring that the random variates are suitable for numerical integration and random simulation, in that the variates are as independent as possible, while the rejection rate is within acceptable bounds.

It is important to note that although in theory the convergence of the sequence to the required distribution is independent of the proposal density, it is the choice of proposal density that is the prime factor in determining the *rate of convergence* of the chain to its equilibrium distribution. This can be easily seen if one considers equation 4.21. If the proposal density $s(\zeta)$ has a large standard deviation relative to $p(X)$, then for most Y_i we will have $p(Y_i) \ll p(X_i)$. Most proposed changes in value are rejected so the random walk is piecewise constant with some large abrupt changes. If, on the other hand, the proposal density $s(\zeta)$ has a small standard deviation relative to $p(X)$ then we will have $p(Y_i) \approx p(X_i)$. There is approximately a $50 : 50$ chance of a change in the value of the sequence from one element to the next, but changes tend to be small and the random walk is therefore very slowly varying. It can be seen intuitively that the best compromise is to endeavour to equate the standard deviations of $p(X)$ and $s(\zeta)$ as closely as possible. This useful rule of thumb is also discussed by Kalos and Whitlock [59].

Another heuristic that is often used is to choose thick tailed proposal densities. This is consistent with an all or nothing philosophy, in which the majority of sample variates ζ_i in equation 4.18 are modest in magnitude, but that occasionally the algorithm makes an ambitious attempt to make large transitions. This latter feature is particularly useful in simulated annealing [53, 82, 125].

4.5.4 Relationship between Gibbs and Metropolis

An interesting result that is relevant to our discussion is the relationship between the Metropolis algorithm and the Gibbs sampler. The Gibbs sampler is, in fact, a special case of the local Metropolis algorithm [82] in which the base transitions in the Metropolis algorithm affect one component at a time, and where the base transitions are rotated in turn. Assume the joint density is given by $p(A, B)$ and it is proposed to change component B to C. For Gibbs sampling, the proposal density is none other than the conditional density $p(C \mid A)$. Hence the Metropolis acceptance function in equation 4.20 is given by:

$$Q\,(C,\, B) = \frac{p(A,\, B)\, p(C \mid A)}{p(A,\, C)\, p(B \mid A)} = 1 \tag{4.23}$$

In other words, the proposal is always accepted and the transition from B to C is always made.

4.6 Dynamical sampling methods

In this section, we discuss Markov chain sampling methods that derive from the "molecular dynamics" approach of Alder and Wainwright [3] which was developed in parallel with the Metropolis algorithm as a means of simulating physical stochastic systems. The methods described below apply only to systems with continuous state variables and where the posterior density is differentiable.

The stochastic dynamics method is particularly interesting for a number of reasons. Foremost among these is the fact that in addition to using only the magnitude of the posterior density to simulate samples, gradient information is used as well. This is also a negative factor because it does require that gradient information is available, which is not the case for discrete parameter spaces.

The gist of the stochastic dynamics method is as follows. The posterior density is defined by the data model. From the canonical distribution we define a notional "energy function" of the model as minus the natural logarithm of the posterior density. The energy function depends only on the position, and hence it may be regarded as corresponding to the concept of potential energy in physical systems. From a physical point of view having only the potential energy available makes the description of the system incomplete because it does not include the familiar effect of "kinetic energy" on the system. To incorporate this effect we augment the position coordinates $\mathbf{q}$ with a set of momentum coordinates $\mathbf{p}$. The Hamiltonian for the system is then given by the usual sum of potential energy and kinetic energy. We consider the properties of a Newtonian dynamical system, the simulation of which surprisingly gives a rather elegant method of producing samples from the posterior density.

Numerical errors (due to rounding etc.) in simulating the dynamics can be cancelled by embedding the stochastic dynamics approach inside the Metropolis algorithm. This is the basis of the Hybrid Monte Carlo algorithm.

As we will see, using a dynamical approach exhibits one great advantage over the pure Metropolis algorithm in that it avoids random walk behaviour. This is primarily due to the fact that the dynamics make use of the gradient of the energy function (or the "force"), which means in effect that kinetic energy can be exchanged for potential energy in order to change position. This has important implications for simulated annealing as discussed in section 4.7.2.

4.6.1 Derivation

Let us consider the canonical distribution of the real variables $\mathbf{Q} = (Q_1, Q_2, Q_3, , \ldots Q_N)$ with respect to the potential energy function

$$p(\mathbf{q}) = \frac{1}{Z_E} \exp\left[-\frac{E(\mathbf{q})}{T}\right] \tag{4.24}$$

where Z_E is the normalization constant. The physical analogy to the above is that the variables Q_i would be the position coordinates for some molecule being considered. In Bayesian inference these would be the model parameters that we are interested in simulating.

The model parameters (or "position coordinates") are augmented with additional variables $\mathbf{P} = (P_1, P_2, P_3, \ldots P_N)$ where each element P_i corresponds to a single Q_i. These elements are the components of the momentum vector for the molecule under consideration. There is no real interpretation for the momentum in Bayesian inference and the momenta are best considered to be no more than dummy variables that are introduced to transform the sampling problem into one of performing dynamical simulations.

The kinetic energy is defined as:

$$K(\mathbf{p}) = \frac{1}{2}\sum_{i=1}^{N} p_i^2 \tag{4.25}$$

The corresponding distribution for these variables is obviously:

$$p(\mathbf{p}) = \frac{1}{Z_K} \exp\left[-K(\mathbf{p})\right]$$

$$= \quad (2\,\pi)^{-N/2} \exp\left[-\frac{1}{2}\sum_{i=1}^{N} p_i^2\right] \tag{4.26}$$

In other words, the components of the momentum are i.i.d. Gaussian variates of unit variance and zero mean.

The combined space of position and momentum vectors is termed *phase space.*

The total energy function (or Hamiltonian) is simply the sum of potential energy and kinetic energy:

$$H\,(\mathbf{p},\,\mathbf{q}) = E\,(\mathbf{q}) + K\,(\mathbf{p}) \tag{4.27}$$

where the temperature T is assumed equal to one.

The canonical distribution over phase space becomes:

$$\begin{aligned} p\,(\mathbf{q},\,\mathbf{p}) &= \frac{1}{Z_H}\exp\left[-E\,(\mathbf{q}) - K\,(\mathbf{p})\right] \\ &= p\,(\mathbf{q})\,p\,(\mathbf{p}) \end{aligned} \tag{4.28}$$

We wish to simulate $p\,(\mathbf{q})$ which is the marginal distribution of the joint density in equation 4.28. The strategy for generating samples from $p\,(\mathbf{q})$ is to generate samples of $\mathbf{q}$ and $\mathbf{p}$ from $p\,(\mathbf{q},\,\mathbf{p})$ jointly, and simply ignore the momentum variates.

The temperature T is implicit in the variance of the components of the momentum vector $\mathbf{p}$. For sampling the canonical distribution the variance can be set to T.

4.6.2 Hamiltonian dynamics

Neal [82] defines a set of dynamical system equations in phase space which exhibit remarkable properties that give a rather elegant way of generating random variates from the posterior density.

The Newtonian dynamical system equations are defined as follows:

$$\textbf{Velocity:} \qquad \frac{d\,q_i}{d\,\tau} = \frac{\partial\,H}{\partial\,p_i} = p_i$$

$$\textbf{Force:} \qquad \frac{d\,p_i}{d\,\tau} = -\frac{\partial\,H}{\partial\,q_i} = -\frac{\partial\,E}{\partial\,q_i}$$

The time τ is of course continuous.

The Newtonian dynamical system has two important properties:

Conservation of energy: This is a familiar property of Newtonian dynamical systems. From the dynamical system equations we can write:

$$\frac{d\,H}{d\,\tau} = \sum_i \left[\frac{\partial\,H}{\partial\,q_i}\frac{d\,q_i}{d\,\tau} + \frac{\partial\,H}{\partial\,p_i}\frac{d\,p_i}{d\,\tau}\right]$$

$$= \sum_i \left[\frac{\partial H}{\partial q_i} \frac{\partial H}{\partial p_i} - \frac{\partial H}{\partial p_i} \frac{\partial H}{\partial q_i} \right]$$
$$= 0 \tag{4.29}$$

Conservation of phase space volume: This property is far less familiar but is no less important. Let us consider the time derivative of change of an elemental volume $d\mathbf{q}\, d\mathbf{p}$:

$$\frac{d\mathbf{q}\, d\mathbf{p}}{d\tau} = \sum_i \left[\frac{\partial}{\partial q_i} \frac{d q_i}{d\tau} + \frac{\partial}{\partial p_i} \frac{d p_i}{d\tau} \right]$$
$$= \sum_i \left[\frac{\partial^2 H}{\partial q_i \partial p_i} - \frac{\partial^2 H}{\partial p_i \partial q_i} \right]$$
$$= 0 \tag{4.30}$$

Neal [82] notes the relationship between this result and Liouville's theorem [6, 66].

These two properties ensure that the canonical distribution remains invariant as one traces the dynamics through time. The reason why this is so is easy to see. Consider a volume V_0 in phase space containing a quantity of energy H_0 at time t. At time $t + \delta t$ the shape of the volume in phase space will have changed but from the above the total volume and the energy contained within will remain unchanged.

Neal [82] notes that it is insufficient just to trace the dynamics in time in order to simulate the canonical distribution, because it only allows one to sample the posterior through regions of constant energy. This is not a disadvantage for standard thermodynamical applications because the expected energy in the system is a fixed constant in the thermodynamic limit. For more general applications (e.g. Bayesian integration) it is clear that in order to sample the canonical distribution the posterior must be sampled over all energies.

4.6.3 Stochastic transitions

Andersen [4] has suggested sampling from the distribution over the p_i in order to obtain the necessary range of energies to sample the canonical distribution. From equation 4.26 we can see that the momentum may be realised as a vector of independent Gaussian random variates. The corresponding kinetic energy is distributed over the range $[0, \infty]$ with a chi squared distribution [70] with the number of degrees of freedom equal to the dimensionality of the parameter space.

Andersen's approach [4] to sampling the distribution was carried out in three stages:

1. Choose the direction of time (forward or backwards) at random.

2. Given a sample position-momentum vector $\mathbf{p}$, $\mathbf{q}$ trace the dynamics to a new position-momentum vector $\mathbf{p}'$, $\mathbf{q}'$.

3. Replace one component of the momentum vector with a Gaussian random variate. This is known as a *stochastic transition.*

Each component of the momentum may be replaced in a fixed order in step 3. The similarity of this approach to Gibbs sampling is noted by Neal [82].

Neal [82] also notes that there is really no reason why *all* the momenta cannot be replaced altogether. At the other extreme, one can choose to only partially update each component using a scheme such as the following:

$$p'_i = \alpha p_i + \sqrt{1-\alpha^2}\, n_i \tag{4.31}$$

where n_i is a Gaussian random variate and α is between zero and one. It is easy to show that expression 4.31 describes a first order autoregressive process with unit variance. Neal advocates using 4.31 to avoid the random walk aspect that is introduced into the motion by complete replacement of the momenta after each Metropolis step.

4.6.4 Simulating the dynamics

The dynamical system defined in section 4.6.2 cannot, in general, be solved exactly as a function of time. The usual approach, as outlined by Neal [82], is to discretize the time coordinate and to compute the position and momentum approximately using finite differences.

Given $p_i(\tau)$ and $q_i(\tau)$ we wish to generate $p_i(\tau + N\,\epsilon)$ and $q_i(\tau + N\,\epsilon)$ where ϵ is a suitably small interval in time and N is an integer.

The leapfrog discretization [82] is defined as:

$$\begin{aligned} p_i(\tau + \frac{\epsilon}{2}) &= p_i(\tau) - \frac{\epsilon}{2}\frac{\partial E}{\partial q_i}\, q_i(\tau) \\ q_i(\tau + \epsilon) &= q_i(\tau) + \epsilon\, p_i(\tau) \\ p_i(\tau + \epsilon) &= p_i(\tau + \frac{\epsilon}{2}) - \frac{\epsilon}{2}\frac{\partial E}{\partial q_i}\, q_i(\tau + \epsilon) \end{aligned}$$

The leapfrog steps are repeated N times until the desired quantities are calculated. It is possible to merge the last step of one stage with the first step of the following stage to form a single step. This explains the name "leapfrog" because the q_i are computed at times $(\tau,\ \tau + \epsilon,\ \tau + 2\,\epsilon,\ \tau + 3\,\epsilon,\ \ldots,\ \tau + N\,\epsilon)$ and the p_i are computed at times $(\tau,\ \tau + \epsilon/2,\ \tau + 3\,\epsilon/2,\ \tau + 5\,\epsilon/2,\ \ldots\ \tau + (2\,N - 1)/(2\,N)\epsilon,\ \tau + N\,\epsilon)$, so except for the very first and the very last time steps the momenta are only computed for times half way across an interval.

Neal [82] shows that the expected error in the simulated values for p_i and q_i in each stage is of order $\mathcal{O}(\epsilon^3)$. The accumulated error after N such stages is of order $\mathcal{O}(\epsilon^2)$. The error in the total energy H is of the same order. There is therefore some systematic error in the final values for p_i and q_i. The bias is reduced by taking more steps for a given length of trajectory but it cannot be entirely eliminated. This places a limit on the number of stages N of the leapfrog discretization that one can reliably carry out for a given step size.

4.6.5 Hybrid Monte Carlo

The Hybrid Monte Carlo algorithm [31, 82] combines stochastic dynamics with the Metropolis algorithm. The stages are as follows:

Direction of time: Choose the direction of time (forward or backwards) at random.

Dynamical simulation: Given a sample position-momentum $\mathbf{p}(\tau)$, $\mathbf{q}(\tau)$ vector trace the dynamics to a new position-momentum vector $\mathbf{p}' = \mathbf{p}\,(\tau + N\,\epsilon)$, $\mathbf{q}' = \mathbf{q}\,(\tau + N\,\epsilon)$. The leapfrog discretization may be used to do this.

Metropolis step: Suppose the total energy is H before performing N leapfrog steps and that it is H' afterwards. A Metropolis decision is made whether to move to the new position ($\mathbf{q}'$, $\mathbf{p}'$) or to stay at the old position ($\mathbf{q}$, $\mathbf{p}$). The decision to move is made with probability $\min(1, \exp[-(H' - H)])$.

Stochastic transition: Replace one or all of the components of the momentum either completely or partially by a Gaussian random variate.

The effect of the Metropolis step is to completely cancel out the effects of bias due to the inaccurate simulation of the dynamical system above. If the dynamical simulation were accurate then we would have $H = H'$ and the new position $\mathbf{q}'$, $\mathbf{p}'$ will always be accepted. Setting $N = 1$ gives a special case known as the Langevin Monte Carlo method.

Neal [82] discusses the potential wastefulness of the Hybrid Monte Carlo scheme; after computing a long trajectory it seems rather wasteful to reject an entire trajectory because of the end position. Neal's solution is to use windowed Monte Carlo: the decision is made whether to proceed to an accept "window" or remain in a reject "window" based on the sum of the probabilities in each. Whether a particular position is accepted is based on weighted sampling from within the window according to the probabilities of each position computed along a trajectory. This scheme is useful if the rejection rate for long trajectories is rather high.

4.6.6 Convergence to canonical distribution

The dynamical system element of the algorithm simulates the canonical distribution, almost by definition. The Metropolis steps are there simply to correct for systematic errors in simulating the dynamics. It is easy to show as in appendix F that the time reversible Hybrid Monte Carlo algorithm satisfies detailed balance.

The simplest way of regarding the Hybrid Monte Carlo algorithm is as a global Metropolis algorithm within which the step of drawing the next sample from the proposal density is replaced by tracing the dynamics using several leapfrog steps.

4.7 Implementation of simulated annealing

Simulated annealing was introduced in section 3.2. There are two elements to the simulated annealing algorithm:

Sampling the canonical distribution: Some means must be found to sample the canonical distribution. In some limited cases the sampling may be done directly. For instance, if the canonical distribution is Gaussian then the temperature acts as a scaling factor for the variance. In general a Markov chain approach to sampling is required. The most common solution is to use the Metropolis algorithm. However there is no reason why one cannot use Gibbs sampling or Hybrid Monte Carlo instead.

Annealing schedules: The selection of a suitable annealing schedule is crucial to the success of simulated annealing. Two classes of annealing schedule, namely predetermined schedules and adaptive schedules, are described below.

In this section, we describe Markov chain approaches to sampling the canonical distribution as well as commonly encountered annealing schedules.

4.7.1 Annealing schedules

The principle that may be gleaned from the discussion about annealing in section 3.2 is that the lowest energy states can only be obtained if the rate of temperature decrease is sufficiently slow, or equivalently, if the annealing schedule is appropriate to the given situation. The analogous situation from our perspective, where we are attempting to minimize the energy function, is that we initialize the search at a high temperature and gradually cool towards absolute zero as the search progresses.

Predetermined schedules

The most well known annealing schedule is the *logarithmic* schedule where the temperature is reduced according to the scheme:

$$T_k = \frac{T_0}{\log(k)} \tag{4.32}$$

where k is time and T_0 is the initial temperature. Ingber and Rosen [53], Geman and Geman [36] and Neal [82] all demonstrate that this annealing schedule *guarantees* convergence to the global minimum of the energy function, providing the initial temperature is greater than the difference between the maximum and minimum energies in the system.

Another scheme that is very common is the *geometric* annealing schedule which is defined by the relation:

$$T_k = \alpha^k T_0 \qquad 0 < \alpha < 1 \tag{4.33}$$

The initial temperature and the value for α are usually set by experimentation.

The logarithmic annealing schedule is not very practical for many problems because of its slow convergence[7]. The geometric annealing schedule is more often used even though it is not always guaranteed to converge to the global minimum.

Ingber and Rosen [53] use an *exponential* annealing schedule of the form

$$T_k = T_0 \exp(-c\, k^{1/D}) \tag{4.34}$$

where D is the dimensionality of the parameter space. They show that convergence to the global minimum is guaranteed if the proposal density used in the Metropolis algorithm in equation 4.18 is chosen carefully. Proposed changes in state ζ_i in equation 4.18 are generated using their own choice of proposal density which is designed to sample over a closed interval. The constant in equation 4.34 is $c = m \exp(-n/D)$ where m and n are constants that are problem dependent.

The choice of annealing schedule that one makes will always depend on the problem to be solved and on practical aspects such as how many function evaluations are acceptable. In this regard some experimentation is always required to test an annealing schedule before it is used in earnest, to set parameters such as the annealing schedule, the initial temperature and the stopping temperature. This point is discussed in more detail by Ingber and Rosen [53] and Press *et al* [95].

[7] Neal [82] notes that this schedule is in fact almost an exhaustive search.

Adaptive Schedules

Adaptive annealing schedules select the temperature for the next iteration from observed characteristics of the system during previous iterations. The objective is to anneal slowly at those critical temperatures where the main features of the low-temperature state become established [82].

One approach developed by Otten and van Ginniken [88] is to strive for a constant rate of *entropy reduction*. For the canonical distribution, entropy is a function of temperature. It can be shown that:

$$\frac{dS}{dT} = \frac{C}{T} \tag{4.35}$$

where C is the heat capacity, which is the rate of change of the expected energy with temperature. The annealing schedule is therefore:

$$T_{k+1} - T_k \propto \frac{T}{C} = \frac{T_k^3}{\langle E^2 \rangle_k - \langle E \rangle_k^2} \tag{4.36}$$

As noted by Neal [82] the need to estimate expectations requires that one keeps the system at fixed temperature, so the annealing schedule will in fact be piecewise constant.

Salamon *et al* [110] also propose an annealing schedule based on constant thermodynamic speed. In this scheme, the system remains at a constant distance from equilibrium at each stage, where distance is measured in terms of the difference between mean energy and the equilibrium energy distribution.

Ingber and Rosen [53] propose re-annealing after every hundred or so acceptances of candidate states, thus resetting the search parameters in the algorithm.

It is worth noting that it is usually extremely difficult to prove theoretically that the global minimum of the energy function will be found when using an adaptive schedule.

4.7.2 Annealing with Markov chains

It is not always easy to sample the canonical distribution directly so we will often need to resort to using Markov chain based techniques to do so.

Metropolis sampling of the canonical distribution

The Metropolis algorithm is very much the standard approach to sampling the canonical distribution, as described by Kirkpatrick [63] and Kirkpatrick, Gellatt and Vecchi [64] and Press *et al* [95]. The most important aspect of applying the Metropolis algorithm is to ensure that the proposal density matches the canonical distribution. In section 4.5.3 it is suggested that the variance of the proposal density match approximately the variance of the distribution being sampled. The effect of temperature on the variance of the canonical distribution is easy to see. If the canonical distribution were Gaussian (i.e. the energy function is quadratic) then the variance would be directly proportional to the temperature. Therefore, as noted by Neal [82], the changes of state in equation 4.18 should be made proportional to the square root of the temperature $\sqrt{T}$.

The proposal distribution affects the annealing schedule. The evidence in the literature, such as the work by Ingber and Rosen [53] and Szu and Hartley [125], suggests that annealing schedules can be considerably shortened if heavy tailed proposal densities are used.

If the conditional densities are of a simple form then one should consider drawing samples from the canonical distribution using a Gibbs sampler. If only some of the conditional densities are of a simple form then a hybrid strategy using combinations of Gibbs and Metropolis stages may be employed.

Hybrid Monte Carlo sampling of the canonical distribution

As noted earlier, the great disadvantage of using random walks to explore parameter space is that the expected time taken to diffuse a certain distance from the starting point varies with the square of the distance. The Hybrid Monte Carlo algorithm moves through state space by a completely different mechanism, using both the magnitude of the canonical distribution and the gradient to make transitions from one state to another. This additional gradient information gives the potential to perform a more direct search than can be achieved by using a random walk alone. An example of the application of Hybrid Monte Carlo to optimization is given by Tidor [129].

The implementation of simulated annealing using Hybrid Monte Carlo is not difficult. One possible implementation is to divide the potential energy function in the canonical distribution by the temperature T. Another possibility, mentioned earlier, is to equate the variance of the momenta to the temperature T. In other words the innovations n_i in equation 4.31 are no longer sampled from a unit variance Gaussian but from a Gaussian density of variance T. The temperature is then controlled externally in the usual way by the annealing schedule.

Neal [82] describes an extremely interesting approach to annealing, based on equation 4.31, that commends itself by its natural simplicity. The momenta are initially chosen to be equal to the square root of temperature $\sqrt{T_0}$. This implies that the expected[8] initial kinetic energy is $T_0\, d/2$ where d is the dimensionality. The innovation sequence in equation 4.31 is subsequently sampled from a unit variance Gaussian. By choosing α just less than 1, there is a "slow forgetting factor" built into the system dynamics whereby one can ensure that the system forgets about the initial high tem-

[8]The kinetic energy is chi squared distributed.

perature slowly because the kinetic energy is dissipated at a suitably slow rate. We define the effective temperature[9] of the system as:

$$T_{\text{eff}} = \frac{2\,K\,(\mathbf{p})}{\mathbf{d}} \tag{4.37}$$

When the effective temperature equals unity the posterior density is being sampled.

The presence of a high kinetic energy term allows the system to overcome high energy barriers along the trajectory. As the system dynamics overcome such a barrier the kinetic energy will decrease, thus decreasing the effective temperature. If the starting point is at a position of high energy then subsequent iterations may actually *increase* the temperature as the system settles towards a position of lower potential energy. This increase in temperature is insurance against the short term option of the dynamics settling into a favourable local minimum in the potential energy function. The opposite will occur if the starting position is of relatively low energy.

4.8 Other issues

Long runs versus short runs

One issue that is of considerable importance when using Markov chains is whether it is better to run a number of Markov chains on the same problem in parallel or to do a single long run. Short runs in parallel are recommended by Gelfand and Smith [35] in the belief that this produces sequences that are as far as possible independent, and that observing the independence between runs may be used as a means of determining convergence towards equilibrium. Tierney [130] advocates using a single long run for the two simple reasons that it is more reliable to judge where equilibrium has been obtained from a long run then a short run and, more importantly, a long run is more efficient because only the first M states (the initial "burn in") are discarded, whereas in N separate runs no less than MN states must be discarded.

Neal [82] recommends using a small number (approximately ten) medium length runs. His belief is that one really should only use a single run if there is reasonable doubt whether the Markov chain will converge in the number of iterations available.

The proposal by Gelfand and Smith [35] to use a large number of relatively short runs is really only suitable for Markov chains that converge relatively quickly, or where one is computing the expectation of a function that is expensive to compute, so the overheads involved in computing the states of the Markov chain are not very significant.

---

[9]This is only a crude analogy to physical temperature.

4.8.1 *Assessing convergence of Markov chains*

According to Tierney [130], it is important while using Markov chains to monitor the performance of the sampler to ensure that the behaviour is as expected. Identifying the convergence of a Markov chain is a difficult problem; several authors (Neal [82], Tierney [130], Tanner [126]) make the point that convergence is a property of the chain rather than of any particular realisation. The methods used to assess convergence of a chain are therefore based on heuristics.

Hypothesis tests for convergence have been proposed by many authors; Ripley [104] gives a review of a number of such methods. A particularly good overview of effective methods for determining convergence is given by Tanner [126].

Ritter and Tanner [106] and Tanner [126] have proposed an approach known as the Gibbs stopper. Let us assume that the transition probabilities may be computed. If all previous runs are regarded as having achieved equilibrium, then sample variates from the chain will be drawn from the equilibrium distribution. The invariant distribution in equation 4.3 may be estimated using importance sampling over previous runs:

$$\pi(x) = \sum_{i=1}^{M} T(x, x_i) \tag{4.38}$$

The following ratio compares estimates of the value of the equilibrium distribution:

$$w_j = f(x_j)/\pi(x_j) \tag{4.39}$$

where $f(x)$ is the probability density being simulated. A near constant value indicates convergence. The above may be modified to monitor convergence in a single long run by cutting the long run into sections or "batches".

Tanner [126] also discusses the use of control variates to monitor convergence, in which expectations of various different functions are calculated using the sample variates. Statistical variations due to the sample realisation are compensated for by using control variates, and an overall diagnostic function is computed, a constant value of which indicates convergence.

4.8.2 *Determining the variance of estimates*

A Markov chain may be used to sample from a predetermined distribution, and these samples may be used to estimate the expected values of functions. Finding a reliable estimate of error in the expectation is complicated by the presence of correlations between successive samples. Neal [82] notes that using equation 3.69 tends to give an over optimistic estimate of the true error. All techniques that attempt to compute variance are therefore designed to compensate for, or strive to remove, correlations between successive samples used to estimate the variance.

The simplest approach is *batching* [82], in which the chain is divided up into batches and the variance computed from the batch means. There are also more sophisticated approaches [82] based on *time series* techniques. Time domain approaches include autocovariance estimates to determine the variance. Frequency domain methods flatten the power spectrum and estimate the variance from the spectral density at zero frequency.

4.9 Free energy estimation

Several different techniques exist in the statistical mechanics literature for the computation of free energy. In section 4.3.3 we noted the equivalence of free energy with Bayesian model evidence. In this section we review some of the techniques currently used in statistical mechanics for the computation of free energy, with a view to applying them to computing Bayesian evidence for model selection.

4.9.1 Thermodynamic integration

Thermodynamic integration [82] is a Markov chain importance sampling approach for the computation of the relative evidence between two models.

Consider two models with partition functions (i.e. Bayesian evidence) Z_1 and Z_2 which may be defined as follows:

$$Z_1 = \int f_1(\mathbf{x})\, d\mathbf{x} \tag{4.40}$$

$$Z_2 = \int f_2(\mathbf{x})\, d\mathbf{x} \tag{4.41}$$

Let the energy functions be defined in the usual manner as

$$E_1 = -\log\left[f_1(\mathbf{x})\right] \tag{4.42}$$

$$E_2 = -\log\left[f_2(\mathbf{x})\right] \tag{4.43}$$

We shall consider the properties of an *intermediate system* with energy function

$$E_\alpha = \alpha\, E_1 + (1-\alpha)\, E_2 \tag{4.44}$$

where $0 < \alpha < 1$. If $\alpha = 1$ then the system has energy function $E_1(\mathbf{x})$ and if $\alpha = 0$ then the system has energy function $E_2(\mathbf{x})$. Let Z_α be the partition function of the intermediate system with energy function $E_\alpha(\mathbf{x})$.

It is easy to show that:

$$\frac{\partial \log Z_\alpha}{\partial \alpha} = -\frac{1}{Z_\alpha}\int \frac{\partial E_\alpha}{\partial \alpha} \exp\left[-E_\alpha\right] d\mathbf{x} \tag{4.45}$$

From equation 4.44 we have:

$$\frac{\partial E_\alpha}{\partial \alpha} = E_1 - E_2 \tag{4.46}$$

Moreover we can factor out a term in equation 4.45

$$f_\alpha(\mathbf{x}) = \frac{1}{Z_\alpha} \exp[-E_\alpha] \tag{4.47}$$

as a normalized probability density.

The term $\frac{\partial}{\partial \alpha} \log Z_\alpha$ can be evaluated by Monte Carlo integration as follows:

$$\begin{aligned} \frac{\partial \log Z_\alpha}{\partial \alpha} &= - \int (E_1(\mathbf{x}) - E_2(\mathbf{x}))\, f_\alpha(\mathbf{x})\, d\mathbf{x} \\ &\approx \tfrac{1}{N} \textstyle\sum_{i=1}^{N} E_1(\mathbf{x}_i) - E_2(\mathbf{x}_i) \end{aligned}$$

where samples $\mathbf{x}_i$ are drawn from the probability density $f_\alpha(\mathbf{x})$. This can easily be carried out using the Metropolis algorithm.

The key to thermodynamic integration is the *slow growth* method of Bash *et al* [9]. We have

$$\log\left[\frac{Z_2}{Z_1}\right] = - \int_0^1 \frac{\partial \log Z_\alpha}{\partial \alpha}\, d\alpha \tag{4.48}$$

This integral may be evaluated by varying α *very* slowly from zero to one while sampling from $f_\alpha(\mathbf{x})$ using Metropolis sampling and evaluating $E_1 - E_2$ at the sample points. The log relative evidence then becomes the sum of the differences $E_1 - E_2$. The number of Metropolis steps that are used is not critical—in fact Neal [82] states that a single Metropolis step for each change in α should be sufficient, providing of course that α is changed slowly enough, so that samples of $E_1 - E_2$ are of approximately uniform size.

As noted by Neal [82] the slow growth method may only be used to evaluate the relative evidence between any two models that are defined over the *same* parameter space. It is possible to get around this problem in practice by pooling the parameters of both systems.

In our case we shall actually concentrate on computing the *absolute* evidence of a model. To do this all we need to do is compare the model with a second system for which the evidence is precisely known. We shall accomplish this by defining a second system which is a Gaussian approximation to the actual system under consideration. This choice of comparison system also has the advantage that it is usually sufficiently close to the true system that few steps are required when changing α to mutate the true system into the dummy system.

4.9.2 Other methods

Importance sampling as described in section 3.8 is an obvious approach to computing free energy. This approach is only useful however if a good importance sampling function can be found.

Neal [82] describes the *acceptance ratio* method of computing free energy differences between distributions. This approach calculates the proportion of one distribution that overlaps with another to compute free energy difference (and hence relative energy). Many variants of the acceptance ratio method are available, including the interpolation method [10] and umbrella sampling [135].

4.10 Summary

In this chapter Markov chain Monte Carlo (MC^2) and stochastic dynamical methods were reviewed. These methods provide a flexible approach to random sampling and have yielded useful techniques for the optimization, integration and simulation of posterior densities.

The optimization of multimodal densities may be carried out using simulated annealing. As we have seen, simulated annealing requires both the specification of an annealing schedule and some means of sampling from the canonical distribution. The standard approach to sampling the canonical distribution is to use the Metropolis algorithm. Another approach, which has great potential, is the Hybrid Monte Carlo algorithm.

Integration by Monte Carlo methods is also facilitated by the use of Markov chains. The sampling techniques in this chapter generate samples from the joint density directly, which is the optimum density for Monte Carlo importance sampling. The Metropolis algorithm and Gibbs sampler are already in common use for this purpose. There is no reason why the Hybrid Monte Carlo algorithm should not also be used; indeed it will be used for this purpose in the subsequent experimental chapters.

A relationship between energy function and probability density was proposed based on the canonical distribution, and was developed in the stochastic dynamics method. We defined a Hamiltonian and introduced a hypothetical time coordinate. From this, a system of dynamical equations was defined under which the Hamiltonian remains constant in time and phase space volume is conserved. This yielded a very elegant and powerful method for simulating the posterior density.

Techniques used for the estimation of free energy in statistical mechanics are also useful for computing Bayesian evidence. A technique known as thermodynamic integration was discussed in some detail.

5

Retrospective Changepoint Detection

5.1 Introduction

The problem of detecting and estimating the location of changepoints (or discontinuities) in data is fundamental to many areas of data analysis. Practical applications abound in diverse areas such as medicine (e.g. monitoring of drug levels in hospital patients), the detection of kickback in oil well pressure data [140] and edge detection in images [123]. In this chapter, optimal Bayesian techniques are developed for changepoint identification in one dimensional (time series) data. These use the probability density function (pdf) of the changepoint positions to estimate their positions in time series.

The chapter begins by examining a very important example, namely the location of the change of mean in a Gaussian process. The general piecewise linear (GPL) model is introduced as a compact matrix formulation of the changepoint problem. The GPL model formulation allows one to easily expand the scope of analysis to the identification of multiple changepoints in more complex data models. This includes piecewise polynomial modelled data (such as cubic splines) and autoregressive (AR) modelled data.

The matrix computations for the GPL model can be very expensive. This is especially true for long data sequences and large model orders. Recursive approaches based on the Woodbury formula [95] are devised for computing the pdf of the location of a candidate changepoint from the pdf at a neighbouring changepoint. The recursive approach often gives enormous savings in computation.

The pdf of the changepoint positions is not usually a very simple function to optimize. In a typical data sequence the pdf for the changepoints will contain a large number of local maxima in addition to a relatively large global maximum. The pdf is very expensive to compute and the parameter space is often large and complicated. This means that standard search algorithms are not very appropriate for this kind of problem. In this chapter, simulated annealing is used as a robust and relatively efficient method for obtaining a near optimal solution.

Using these methods the GPL model is extended to study multiple changepoints in non-Gaussian impulsive noise environments. This is used to analyse well log data to determine the boundaries between different rock formations.

5.2 The simple Bayesian step detector

5.2.1 Derivation of the step detector

In this section, it is assumed that there are N samples of data from a piecewise constant input with added Gaussian noise:

$$d_i = \begin{cases} \mu_1 + e_i & \text{if } i < m \\ \mu_2 + e_i & \text{otherwise} \end{cases} \tag{5.1}$$

The noise samples e_i are assumed independent.

The likelihood of the data is given by the joint probability of the noise samples.

$$p(\mathbf{d} \mid \{\mu_1\mu_2\sigma m\}, \mathbf{I}) \quad = \textstyle\prod_{i=1}^{N} p(e_i) \tag{5.2}$$

where σ is the standard deviation of the Gaussian noise, $\{\mu_1\mu_2\sigma m\}$ denotes various assumed values of the signal parameters and $\mathbf{I}$ denotes prior information, or any assumptions which led to this particular choice of signal model.

Substituting equation 5.1 into equation 5.2 leads to the following expression for the likelihood:

$$p(\mathbf{d} \mid \{\mu_1\mu_2\sigma m\}, \mathbf{I}) = \left(2\pi\sigma^2\right)^{-\frac{N}{2}} \exp\left[-\frac{1}{2\sigma^2}\left(\sum_{i=1}^{M} e_i^2\right)\right] \tag{5.3}$$

where the energy of the residuals is given by:

$$\sum_{i=1}^{M} e_i^2 = \sum_{i=1}^{m} (d_i - \mu_1)^2 + \sum_{i=m+1}^{N} (d_i - \mu_2)^2 \tag{5.4}$$

At this point, parameter estimates could be obtained for the three continuous parameters (μ_1, μ_2, σ) by the method of maximum likelihood (ML). Differentiating equation 5.3 with respect to each of μ_1, μ_2, σ and equating to zero and solving yields closed form expressions for the supremum $(\hat{\mu}_1, \hat{\mu}_2, \hat{\sigma})$. The value of the discrete parameter m can then be inferred by back-substitution of the ML parameter estimates into the likelihood and maximizing the resultant expression. This is the basis of the Hinckley test [118] which is a common approach for finding changepoints.

The Bayesian approach relies on Bayes' theorem which for our purposes may be stated as follows:

$$p(\{\mu_1\mu_2\sigma m\} \mid \mathbf{d}, \mathbf{I}) = \frac{p(\mathbf{d} \mid \{\mu_1\mu_2\sigma m\}, \mathbf{I}), p(\{\mu_1\mu_2\sigma m\} \mid \mathbf{I})}{p(\mathbf{d} \mid \mathbf{I})} \tag{5.5}$$

The term in $p(\{\mu_1\mu_2\sigma m\} \mid \mathbf{I})$ is the prior probability density for the parameters, which when multiplied by the likelihood is *updated* to give the posterior probability density (of the parameters given the data). The posterior density is used to infer the values of parameters given the observed data realization $\mathbf{d}$.

In this case, non-informative prior probability densities are chosen to express ignorance about the value of the parameter vector in the absence of data. If it is assumed that all parameter values are equally likely a priori then:

$$\begin{aligned} p(\mu_1 \mid \mathbf{I}) &= k_1 \\ p(\mu_2 \mid \mathbf{I}) &= k_2 \\ p(\log(\sigma) \mid \mathbf{I}) &= k_3 \Longrightarrow p(\sigma \mid \mathbf{I}) \propto \frac{1}{\sigma} \end{aligned} \tag{5.6}$$

where each k_i is a constant.

Substituting equation 5.6 into equation 5.5 yields

$$p(\{\mu_1\mu_2\sigma m\} \mid \mathbf{d}, \mathbf{I}) \propto \frac{1}{\sigma}\, p(\mathbf{d} \mid \{\mu_1\mu_2\sigma m\}, \mathbf{I}) \tag{5.7}$$

The next stage in the process of Bayesian inference is to remove the so called "nuisance parameters" by integration.

$$p(\{m\} \mid \mathbf{d}, \mathbf{I}) = \int_0^{\infty} d\sigma \int_{-\infty}^{\infty} d\mu_1 \int_{-\infty}^{\infty} d\mu_2\, p(\{\mu_1\mu_2\sigma m\} \mid \mathbf{d}, \mathbf{I}) \tag{5.8}$$

Two standard integral identities are of use in computing expression 5.8. First, there is the integral identity:

$$\int_{-\infty}^{\infty} \exp\left(-a\,x^2 - b\,x - c\right) dx = \sqrt{\frac{\pi}{a}} \exp\left(\frac{b^2}{4a} - c\right) \tag{5.9}$$

Second, the gamma integral in expression 2.27 is used:

$$\int_{0}^{\infty} x^{\alpha-1} \exp(-Q\,x)\,dx = \frac{\Gamma(\alpha)}{Q^{\alpha}} \tag{5.10}$$

The first mean μ_1 may be integrated out using expression 5.9:

$$p(\{\mu_2 \sigma m\} \mid \mathbf{d},\, \mathbf{I}) = \int_{-\infty}^{\infty} p(\{\mu_1 \mu_2 \sigma m\} \mid \mathbf{d},\, \mathbf{I})\, d\mu_1$$

$$\propto \frac{\left(2\,\pi\sigma^2\right)^{-\frac{N}{2}}}{\sqrt{m}} \exp\left[-\frac{1}{2\,\sigma^2}\left(\sum_{i=1}^{m} {d_i}^2 - \frac{1}{m}\,{S_l}^2 + \sum_{i=m+1}^{N} (d_i - \mu_2)^2\right)\right]$$

where

$$S_l = \sum_{i=1}^{m} d_i \tag{5.11}$$

Similarly, the second mean μ_2 may also be integrated out to obtain:

$$p(\{\sigma m\} \mid \mathbf{d},\, \mathbf{I}) = \int_{-\infty}^{\infty} p(\{\mu_2 \sigma m\} \mid \mathbf{d},\, \mathbf{I})\, d\mu_2$$

$$\propto \frac{\left(2\,\pi\sigma^2\right)^{-\left(\frac{N-1}{2}\right)}}{\sqrt{m\,(N-m)}} \exp\left[-\frac{1}{2\sigma^2}\left(\sum_{i=1}^{N} {d_i}^2 - \frac{1}{m}\,{S_l}^2 - \frac{1}{N-m}\,{S_r}^2\right)\right]$$

where

$$S_r = \sum_{i=m+1}^{N} d_i \tag{5.12}$$

Finally, the standard deviation σ is integrated out using the gamma integral in expression 5.10:

$$p(\{m\} \mid \mathbf{d},\, \mathbf{I}) = \int_{-\infty}^{\infty} p(\{m\},\, \sigma \mid \mathbf{d},\, \mathbf{I})\, d\sigma$$

$$\propto \frac{1}{\sqrt{m\,(N-m)}} \left[\sum_{i=1}^{N} {d_i}^2 - \frac{1}{m}\,{S_l}^2 - \frac{1}{N-m}\,{S_r}^2\right]^{-\left(\frac{N-2}{2}\right)} \tag{5.13}$$

Note that expression 5.13 is a function of the changepoint position only. This means that there is no need to know about the standard deviation, probable step position or step size in order to detect changepoints. Here the integrals have been done analytically so the dimensionality of the parameter space was reduced by one for each parameter integrated out. This reduction of the dimensionality is a useful feature in this application.

Smith [118] notes that expression 5.13 is virtually identical to the Hinckley test. This is because the Bayesian result, like the Hinckley test, is derived from the likelihood function. In this case, the prior probability exerts little influence on the Bayesian result.

5.2.2 *Application of the step detector*

We now consider the performance of the Bayesian step detector derived above for the detection of a single changepoint in synthetic data. The method is to test each point in the time series as a potential changepoint using expression 5.13.

Consider the time series plotted in figure 5.1. This graph shows an abrupt step in Gaussian noise. The corresponding probability density for step position is given in figure 5.2. The inferred step position, corresponding to the maximum probability density, is $m = 33$.

The data given in figure 5.3 provides a far more difficult test for the Bayesian step detector. A simple examination of the data by eye gives a changepoint at around position $m = 120$ or position $m = 140$.

In this case the probability density is given in figure 5.4. The Bayesian result is close to what one might expect. There is a small peak at position $m = 125$ but the probability density is a maximum at $m = 140$ which corresponds exactly with the position of the changepoint.

Note that the probability density is more diffuse which indicates that the Bayesian estimator is not confident of the changepoint position because of the higher noise level.

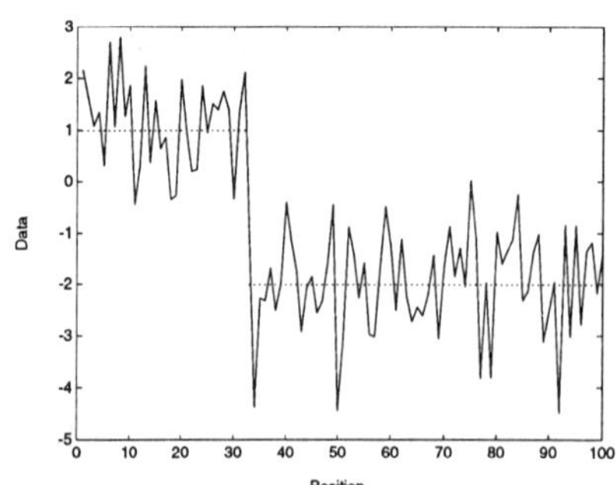

FIGURE 5.1. Discrete step in Gaussian noise

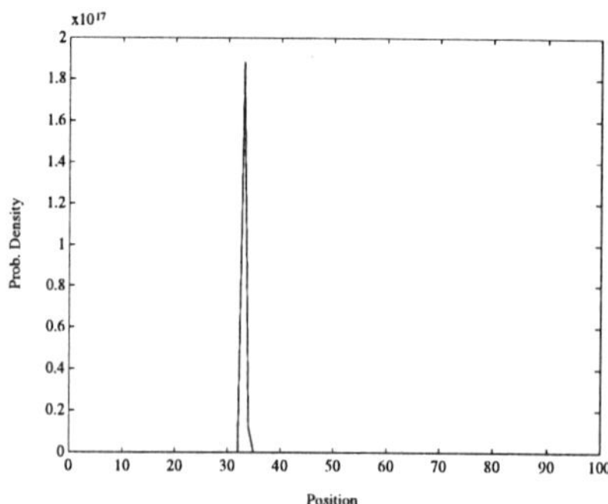

FIGURE 5.2. Bayesian step detection in Gaussian noise

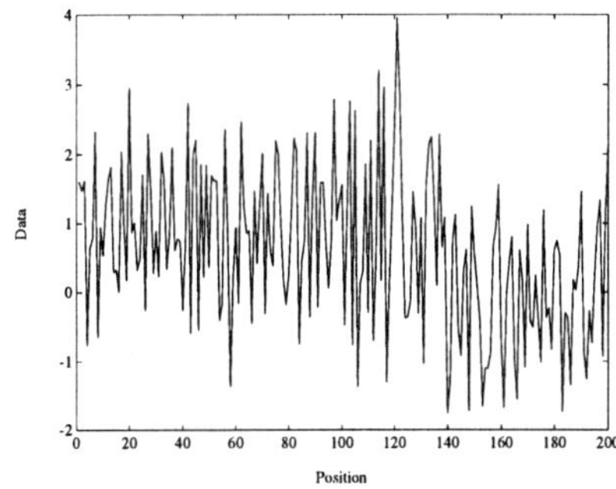

FIGURE 5.3. Discrete step in Gaussian noise

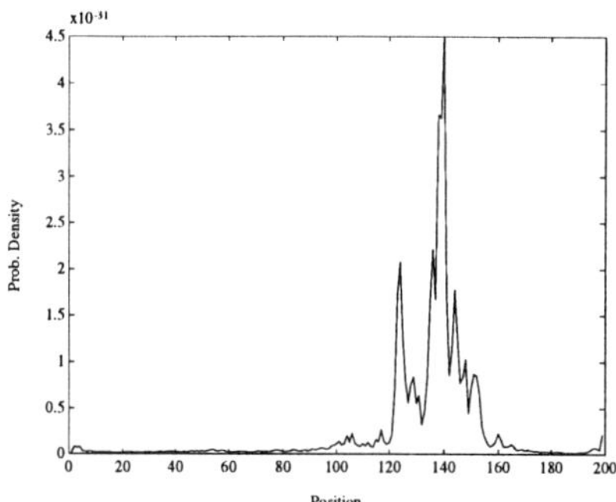

FIGURE 5.4. Bayesian step detection in Gaussian noise

The results are obviously excellent for the particular realization of a Gaussian random process given in figure 5.3. The question arises as to how well the step detector works for other noise realizations. Figure 5.5 is a histogram of the maximum a posteriori estimates obtained over one thousand different noise realizations such as that of figure 5.3. It is clear that there is a very sharp peak at or around the "true" changepoint position $m = 140$.

5.3 The detection of changepoints using the general linear model

In this section a matrix formulation of the signal model is employed, which allows all linear models (such as polynomial fits to data as well as autoregressive models) to be treated in exactly the same way. This model was first developed by Kheradmandnia [62], but was later independently rediscovered [85, 86] in the course of work on changepoint detection.

5.3.1 *The general piecewise linear model*

Suppose the data can be fitted as:

$$y(i) = \begin{cases} \sum_{k=1}^{m} a_k g_k(i) + e(i) & \text{if } i < m \\ \sum_{k=1}^{m} b_k g_k(i) + e(i) & \text{otherwise} \end{cases}$$

where $g_k(i)$ is the value of a time dependent model function $g_k(t)$ evaluated at time t_i.

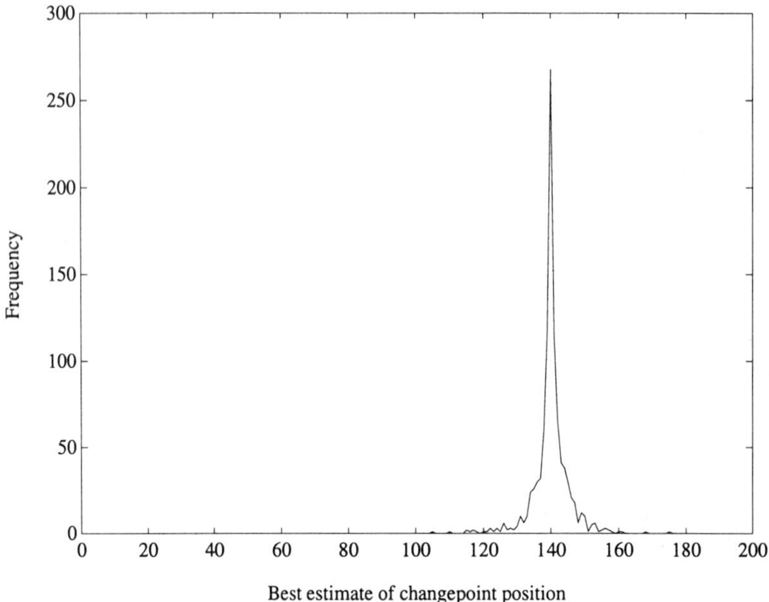

FIGURE 5.5. Performance of step detector

This can be written in the form of a matrix equation

$$\mathbf{d} = \mathbf{G}\,\mathbf{b} + \mathbf{e} \tag{5.14}$$

where $\mathbf{d}$ is an $N \times 1$ matrix of data points and $\mathbf{e}$ is an $N \times 1$ matrix of Gaussian noise samples. The matrix $\mathbf{G}$ is of size $N \times M$. Each column of $\mathbf{G}$ is a basis function evaluated at each point in the time series and each element of the $M \times 1$ matrix $\mathbf{b}$ is a linear coefficient (corresponding to each column of the matrix $\mathbf{G}$). The likelihood function is:

$$p\left(\mathbf{d} \mid \{m\}, \sigma, \mathbf{b}, \mathbf{I}\right) = \left(2\,\pi\,\sigma^2\right)^{-\frac{N}{2}} \exp\left[-\frac{\mathbf{e}^{\mathbf{T}}\mathbf{e}}{2\,\sigma^2}\right] \tag{5.15}$$

where $\{m\}$ denotes the changepoints (non-linear parameters) in the matrix of basis functions $\mathbf{G}$.

Assigning uniform priors to the linear parameters $\mathbf{b}$ and Jeffreys' prior to σ, the nuisance parameters can be integrated out using the methods in appendix A to obtain:

$$p(\{m\} \mid \mathbf{d}, \mathbf{I}) \propto \frac{\left[\mathbf{d}^{\mathbf{T}}\mathbf{d} - \mathbf{d}^{\mathbf{T}}\mathbf{G}\left(\mathbf{G}^{\mathbf{T}}\mathbf{G}\right)^{-1}\mathbf{G}^{\mathbf{T}}\mathbf{d}\right]^{\frac{-(N-M)}{2}}}{\sqrt{\det\left(\mathbf{G}^{\mathbf{T}}\mathbf{G}\right)}} \tag{5.16}$$

The changepoint position is built into the structure of the matrix $\mathbf{G}$.

At this point, the posterior probability of the changepoints $\{m\}$ has been obtained, which can be used to infer their values in the absence of any information regarding the values of the noise level σ or the linear coefficients $\mathbf{b}$ of the basis functions.

5.3.2 Simple step detector in generalized matrix form

In this section, the posterior density is computed for the change of mean for a Gaussian process using equation 5.16.

Consider equation 5.1. The matrix $\mathbf{G}$ will consist of two columns. The first column will consist of m rows of 1s and $N - m$ rows of 0s thereafter. On the other hand, the second column will consist of m rows of 0s and $N - m$ rows of 1s thereafter:

$$\mathbf{G}^{\mathbf{T}} = \begin{bmatrix} 1 & 1 & 1 & 1 & 1 & \cdots & 1 & 0 & \cdots & 0 & 0 \\ 0 & 0 & 0 & 0 & 0 & \cdots & 0 & 1 & \cdots & 1 & 1 \end{bmatrix}$$

This implies:

$$\begin{aligned} \mathbf{G}^{\mathbf{T}}\mathbf{G} &= \begin{bmatrix} m & 0 \\ 0 & N-m \end{bmatrix} \\ \left(\mathbf{G}^{\mathbf{T}}\mathbf{G}\right)^{-1} &= (m(N-m))^{-1}\begin{bmatrix} N-m & 0 \\ 0 & m \end{bmatrix} \\ \mathbf{d}^{\mathbf{T}}\mathbf{G} &= \begin{bmatrix} \sum_{i=1}^{m} d_i & \sum_{i=m+1}^{N} d_i \end{bmatrix} \\ \mathbf{d}^{\mathbf{T}}\mathbf{d} &= \sum_{i=1}^{N} d_i^2 \end{aligned} \tag{5.17}$$

Putting all these terms together, the posterior density becomes:

$$p(\{m\} \mid \mathbf{d}, \mathbf{I}) \propto \frac{\left[\mathbf{d}^{\mathbf{T}}\mathbf{d} - \mathbf{d}^{\mathbf{T}}\mathbf{G}\left(\mathbf{G}^{\mathbf{T}}\mathbf{G}\right)^{-1}\mathbf{G}^{\mathbf{T}}\mathbf{d}\right]^{\frac{-(N-M)}{2}}}{\sqrt{\det\left(\mathbf{G}^{\mathbf{T}}\mathbf{G}\right)}}$$

$$= \alpha \left[\sum_{i=1}^{N} {d_i}^2 - \frac{(N-m)\left(\sum_{i=1}^{M} d_i\right)^2 - m\left(\sum_{i=m+1}^{N} d_i\right)^2}{m(N-m)} \right]^{\frac{-(N-2)}{2}}$$

$$= \alpha \left[\sum_{i=1}^{N} {d_i}^2 - \frac{1}{m}\left(\sum_{i=1}^{m} d_i\right)^2 - \frac{1}{N-m}\left(\sum_{i=m+1}^{N} d_i\right)^2 \right]^{\frac{-(N-2)}{2}} \tag{5.18}$$

where $\alpha = (m(N-m))^{-1/2}$ which is exactly the same as equation 5.13.

5.3.3 Changepoint detection in AR models

In this subsection, equation 5.16 is applied to the detection of changepoints in AR modelled data.

$$y_i = \begin{cases} \sum_{k=1}^{p} a_k y_{i-k} + e_i & \text{if } i < m \\ \sum_{k=1}^{p} \delta_k y_{i-k} + e_i & \text{otherwise} \end{cases} \tag{5.19}$$

Consider a two pole AR model of the form of equation 5.19. This may be expressed in standard form according to equation 5.14:

$$\begin{bmatrix} y_1 \\ y_2 \\ y_3 \\ \vdots \\ y_{m-1} \\ y_m \\ y_{m+1} \\ \vdots \\ y_N \end{bmatrix} = \begin{bmatrix} y_0 & y_{-1} & 0 & 0 \\ y_1 & y_0 & 0 & 0 \\ y_2 & y_1 & 0 & 0 \\ \vdots & \vdots & \vdots & \vdots \\ y_{m-2} & y_{m-3} & 0 & 0 \\ y_{m-1} & y_{m-2} & 0 & 0 \\ 0 & 0 & y_m & y_{m-1} \\ \vdots & \vdots & \vdots & \vdots \\ 0 & 0 & y_{N-1} & y_{N-2} \end{bmatrix} \begin{bmatrix} a_1 \\ a_2 \\ \delta_1 \\ \delta_2 \end{bmatrix} + \begin{bmatrix} e_1 \\ e_2 \\ e_3 \\ \vdots \\ e_{m-1} \\ e_m \\ e_{m+1} \\ \vdots \\ e_N \end{bmatrix} \tag{5.20}$$

From the point of view of the model, the data points y_i where $i < 1$ are observed data points. In other words, one should treat the data as beginning at point $i = 1$ at the third data point. An adhoc alternative is to treat the data as beginning at point $i = 1$ at the first data point and to replace y_i where $i < 1$ by their expected values, namely zero. However, the latter gives inferior results because, although zero is the most probable value, it is certainly not typical. This can cause the detector to indicate a changepoint there instead [62].

The generalization of equation 5.20 above to p pole AR models is obvious. Note that the matrix $\mathbf{G}$ is a function of changepoint position m only. The AR coefficients and the noise standard deviation may be integrated out to give the posterior density in equation 5.16.

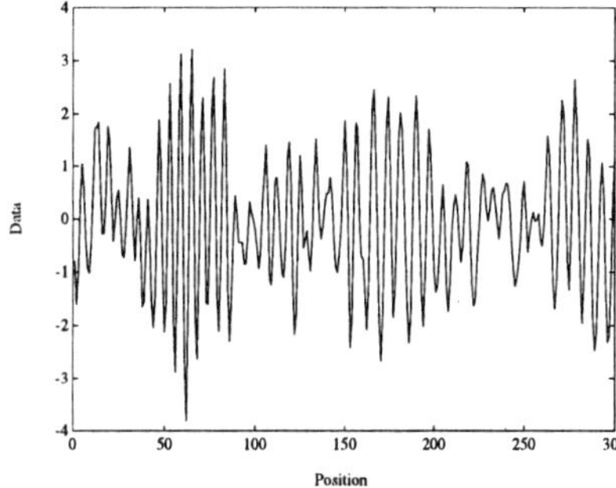

FIGURE 5.6. Three pole AR data with a single changepoint

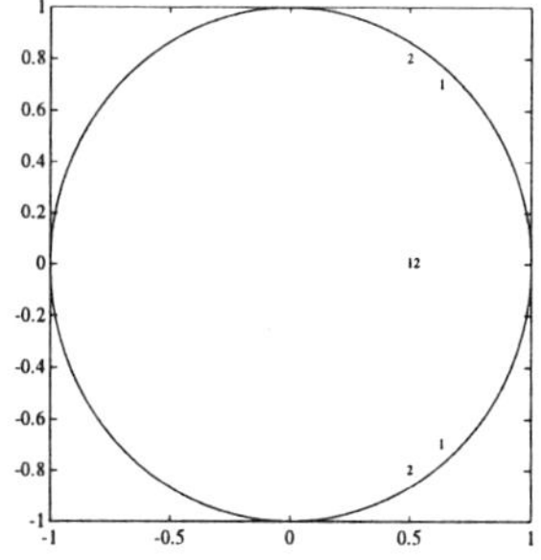

FIGURE 5.7. Pole positions to the left (1) and right (2) of the changepoint

5.3.4 Application of AR changepoint detector

Consider the time series plotted in figure 5.6. The time series is generated by two separate autoregressive processes with an abrupt change in the autoregressive coefficients at position $m = 140$. The poles of both autoregressive processes are plotted in figure 5.7.

The "true" changepoint is at position $m = 140$. The two AR processes on either side of the changepoint are of order three. The pole positions are plotted on the complex plane in figure 5.7.

For the purposes of Bayesian analysis, a three pole AR process is assumed with a single changepoint located somewhere within the data set.

The probability density function for the position of the proposed changepoint is plotted in figure 5.8. Again, the MAP estimate $m = 140$ exactly corresponds to the changepoint position used when generating the data.

The performance of the AR changepoint detector for a large number of trial runs is plotted in figure 5.9. This plot shows the MAP estimate of the changepoint position for one thousand data realizations such as figure 5.6 above. Note that the body of the distribution is well concentrated about the "true" position $m = 140$. However, it is also significant that no changepoints were detected for a certain number of realizations. The null result is indicated by the changepoint being detected at or near the endpoints of the data (so there is little or no evidence to support the hypothesis that a changepoint is present). The model assumes that there is a changepoint, so therefore a changepoint has to be detected somewhere in the data. It is interesting that Bayesian probability theory places this at the endpoints. The reason for this behaviour is the presence of the term in $\det\left(\mathbf{G}^{\mathrm{T}}\mathbf{G}\right)$ which is small close to the endpoints.

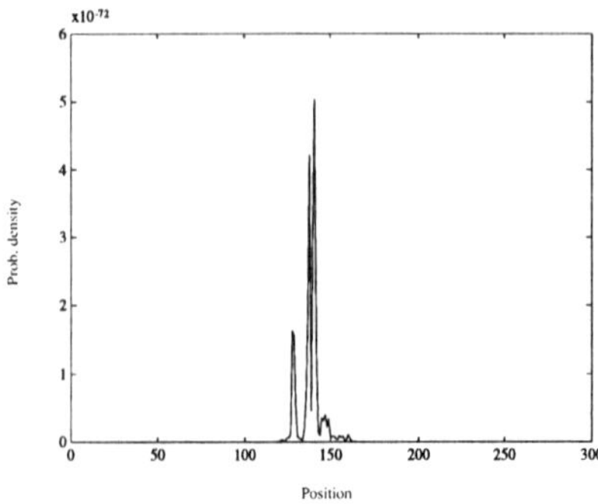

FIGURE 5.8. Bayesian changepoint detection in autoregressive data

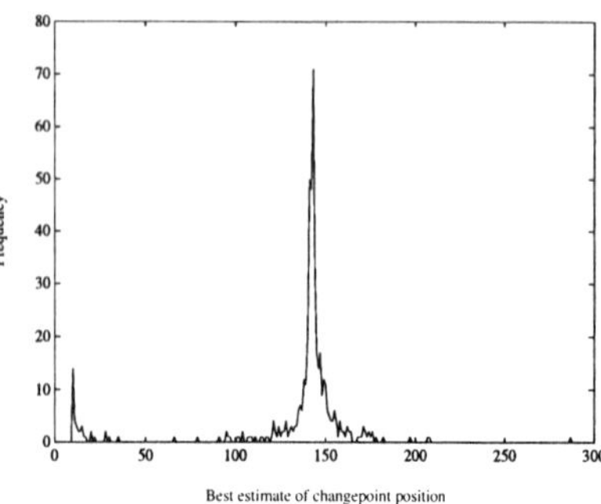

FIGURE 5.9. Histogram of estimates for three pole data

5.4 Recursive Bayesian estimation

As noted by Kheradmandnia [62] and Gordon and Smith [44] direct application of equation 5.16 is difficult in practice because the probability density is extremely expensive to compute. This is especially true when the time series is long. In this section, it is demonstrated how to make the Bayesian changepoint detector recursive. This produces considerable savings in computation and makes the detector efficient enough for practical applications. Two problems will be solved in this section. In the first example, given $p(m \mid \mathbf{d}, \mathbf{I})$ it is shown how to compute $p(m+1 \mid \mathbf{d}, \mathbf{I})$ where m is the current position being tested as a potential changepoint. In the second example, it is also shown how to modify posterior densities to take into account additional data as it arises. Stated more precisely, $p(m \mid \mathbf{d_1}, \mathbf{I})$ will be updated to give $p(m \mid \mathbf{d_1}, \mathbf{d_2}, \mathbf{I})$.

In all there are four terms in equation 5.16 which may be updated recursively. For position update the three terms which need updating are, $\mathbf{G^T d}$, $\left(\mathbf{G^T G}\right)^{-1}$ and $\det\left(\mathbf{G^T G}\right)$. For data update one also needs to consider the term in $\mathbf{d^T d}$.

The greatest difficulty is encountered in attempting to update the term in $\left(\mathbf{G^T G}\right)^{-1}$. The solution depends on the Woodbury formula [95].

$$\left(\mathbf{A}+\mathbf{U}\mathbf{V^T}\right)^{-1} = \mathbf{A}^{-1} - \left[\mathbf{A}^{-1}\mathbf{U}\left(\mathbf{I}+\mathbf{V^T}\mathbf{A}^{-1}\mathbf{U}\right)^{-1}\mathbf{V^T}\mathbf{A}^{-1}\right] \tag{5.21}$$

The update of $\det\left(\mathbf{G^T G}\right)$ is a related problem and is solved using

$$\det\left(\mathbf{A}+\mathbf{U}\mathbf{V^T}\right) = \det\left(\mathbf{A}\right)\det\left(\mathbf{I}+\mathbf{V^T}\mathbf{A}^{-1}\mathbf{U}\right) \tag{5.22}$$

The steps required for the position update and the new data update are detailed in the following two subsections. For both kinds of update it is assumed that the initial matrices which follow are available as required:

$$\mathbf{\Phi} = \left(\mathbf{G^T G}\right)^{-1} \tag{5.23}$$

$$\mathbf{g} = \mathbf{d^T G} \tag{5.24}$$

$$\mathbf{D} = \mathbf{d^T d} \tag{5.25}$$

$$\mathbf{\Delta} = \det\left(\mathbf{G^T G}\right) \tag{5.26}$$

5.4.1 Update of position

Initially, it is assumed that the probability density $p(m \mid \mathbf{d}, \mathbf{I})$ is written in the form:

$$p(m \mid \mathbf{d}, \mathbf{I}) \propto \frac{1}{\sqrt{\mathbf{\Delta}}\left[\mathbf{D} - \mathbf{g_i}\,\mathbf{\Phi_i}\,\mathbf{g_i^T}\right]^{\frac{N-M}{2}}} \tag{5.27}$$

The key matrices $\mathbf{\Delta}$ $\mathbf{g}$ $\mathbf{\Phi}$ and $\mathbf{D}$ will be modified recursively to obtain the density $p(m+1 \mid \mathbf{d}, \mathbf{I})$. The change in posterior probability is often due to a change of a *single* row of the matrix $\mathbf{G}$. Let $\mathbf{R}$ be the old m^{th} row and let $\mathbf{Q}$ be the new row that replaces it. The aim is to update the key matrix terms as follows:

$$\left(\mathbf{G^T G}\right)^{-1} \leftarrow \left(\mathbf{G^T G} - \mathbf{R^T R} + \mathbf{Q^T Q}\right)^{-1}$$

$$\det\left(\mathbf{G^T G}\right) \leftarrow \det\left(\mathbf{G^T G} - \mathbf{R^T R} + \mathbf{Q^T Q}\right)$$

$$\mathbf{d^T G} \leftarrow \mathbf{d^T G} - d_m\mathbf{R} + d_m\mathbf{Q} \tag{5.28}$$

This can be carried out in two stages. The first stage is to replace the mth row of $\mathbf{G}$ with a row of zeroes.

$$\mathbf{g}_{i+1} = \mathbf{g}_i - d_m\mathbf{R}$$

$$\mathbf{W} = \mathbf{\Phi_i R^T}$$

$$\lambda = (\mathbf{1} - \mathbf{RW})$$

$$\mathbf{\Delta_{i+1}} = \lambda\,\mathbf{\Delta_i}$$

$$\mathbf{\Phi_{i+1}}{=}\mathbf{\Phi_i} + \frac{1}{\lambda}\,\mathbf{W}\mathbf{W}^{\mathbf{T}}$$

The second stage is to replace the row of zeros introduced into $\mathbf{G}$ with the matrix $\mathbf{Q}$.

$$\mathbf{g}_{i+2} = \mathbf{g}_{i+1} + d_m\mathbf{Q}$$

$$\mathbf{W} = \mathbf{\Phi_{i+1}}\mathbf{Q}^{\mathbf{T}}$$

$$\mathbf{\Delta_{i+2}} = \lambda\,\mathbf{\Delta_{i+1}}$$

$$\lambda = (\mathbf{1} + \mathbf{Q}\mathbf{W})$$

$$\mathbf{\Phi_{i+2}} = \mathbf{\Phi_{i+1}} - \frac{1}{\lambda}\,\mathbf{W}\mathbf{W}^{\mathbf{T}}$$

Finally, having modified all the relevant terms the posterior density $p\,(m+1 \mid \mathbf{d},\, \mathbf{I})$ is computed as:

$$p\,(m+1 \mid \mathbf{d},\, \mathbf{I}) \propto \frac{1}{\sqrt{\mathbf{\Delta}}\left[\mathbf{D} - \mathbf{g_{i+2}}\,\mathbf{\Phi_{i+2}}\,\mathbf{g}^{\mathbf{T}}_{\mathbf{i+2}}\right]^{\frac{N-M}{2}}} \tag{5.29}$$

in common with expression 5.27.

The matrix $\mathbf{G}^{\mathbf{T}}\mathbf{G}$ has a block diagonal structure (i.e. it is in Jordan form), with each block corresponding to the data between two sets of assumed changepoints. The inverse of a block diagonal matrix is also block diagonal. The inverse of the matrix can be computed by simply inverting each block in turn.

5.4.2 Update given more data

An increase in the number of data leads to an increase in the number of rows in the matrix $\mathbf{G}$. Let $\mathbf{G_1}$ be the model basis matrix corresponding to the original data $\mathbf{d_1}$. If additional data $\mathbf{d_2}$ appears then according to expression 5.14:

$$\begin{bmatrix}\mathbf{d_1}\\ \mathbf{d_2}\end{bmatrix} = \begin{bmatrix}\mathbf{G_1}\\ \mathbf{G_2}\end{bmatrix}\left[\,\mathbf{b}\,\right] + \begin{bmatrix}\mathbf{e_1}\\ \mathbf{e_2}\end{bmatrix} \tag{5.30}$$

The extension of the matrix equations means that the key matrix terms must be updated as follows:

$$\left(\mathbf{G}^{\mathbf{T}}\mathbf{G}\right)^{-\mathbf{1}} \leftarrow \left(\mathbf{G_1}^{\mathbf{T}}\mathbf{G_1} + \mathbf{G_2}^{\mathbf{T}}\mathbf{G_2}\right)^{-\mathbf{1}}$$

$$\det\left(\mathbf{G}^{\mathbf{T}}\mathbf{G}\right) \leftarrow \det\left(\mathbf{G_1}^{\mathbf{T}}\mathbf{G_1} + \mathbf{G_2}^{\mathbf{T}}\mathbf{G_2}\right)$$

$$\mathbf{d}^{\mathbf{T}}\mathbf{G} \leftarrow \left(\mathbf{d_1}^{\mathbf{T}}\mathbf{G_1} + \mathbf{d_2}^{\mathbf{T}}\mathbf{G_2}\right)$$

$$\mathbf{d}^{\mathbf{T}}\mathbf{d} \leftarrow \left(\mathbf{d_1}^{\mathbf{T}}\mathbf{d_1} + \mathbf{d_2}^{\mathbf{T}}\mathbf{d_2}\right) \tag{5.31}$$

Let the key matrix terms be defined as in equations 5.23, 5.24, 5.25 and 5.26. This may be carried out using the following scheme:

$$\mathbf{D_{i+1}} = \mathbf{D_i} + \mathbf{d_2}^{\mathbf{T}}\mathbf{d_2}$$

$$\mathbf{g}_{i+1} = \mathbf{g_i} + \mathbf{d_2}^{\mathbf{T}}\mathbf{G_2}$$

$$\mathbf{W} = \mathbf{\Phi_i}\,\mathbf{G_2}^{\mathbf{T}}$$

$$\mathbf{\Lambda} = (\mathbf{1} + \mathbf{G_2}\mathbf{W})$$

$$\mathbf{\Delta}_{i+1} = \mathbf{\Delta}_i \det\mathbf{\Lambda}$$

$$\mathbf{\Phi_{i+1}} = \mathbf{\Phi_i} - \mathbf{W}\mathbf{\Lambda}^{-1}\mathbf{W}^{\mathbf{T}}$$

Finally, the posterior density $p\left(m \mid \mathbf{d_1}, \mathbf{d_2}, \mathbf{I}\right)$ is given by:

$$p\left(m \mid \mathbf{d_1}, \mathbf{d_2}, \mathbf{I}\right) \propto \frac{1}{\sqrt{\mathbf{\Delta}}\left[\mathbf{D}_{i+1} - g_{i+1}\,\mathbf{\Phi}_{i+1}\,g_{i+1}^{\mathbf{T}}\right]^{\frac{N-M}{2}}} \tag{5.32}$$

where N is the length of the new data set.

5.5 Detection of multiple changepoints

In this section, a sampling based approach is developed for the identification of multiple changepoints in linearly modelled data, where the noise corrupting the signal is of some known distribution and is not necessarily Gaussian. The approach is a generalization of the basic methods used in chapter 2.

It is convenient to partition the parameter space into three components — namely changepoints, amplitudes and noise paremeters:

- **Changepoints:** All positions in the data are assumed to be candidate changepoints. This component of parameter space is discrete. The number of *distinct* possibilities for the position of the changepoints is:

$$C = \frac{N\,!}{(N-d)\,!\,d\,!} \tag{5.33}$$

where d is the number of changepoints and N is the number of data.

The number of possible positions C is exponential in the number of changepoints (i.e. the search problem is *np* complete). This means that it is not feasible to examine all possible changepoint configurations over a multidimensional grid, even for a relatively small number of data.

- **Linear parameters:** It is assumed that linear parameters are real valued and continuous (i.e. $\mathbf{b} \in \Re^p$).

- **Noise parameters:** Noise parameters are usually continuous and bounded on the real axis. For example, the decay of a Laplacian density is strictly positive.

5.6 Implementation details

Consider figure 5.10 which shows the marginal posterior density for two changepoints in Gaussian noise for a typical data set. The posterior density is symmetric because to infer a changepoint at position (m_1, m_2) is obviously the same as to infer a changepoint at (m_2, m_1).

The surface of the posterior density is not particularly simple to optimize. In general, in addition to the dominant mode there are countless small modes. The multimodal nature of the cost function and the huge size of the parameter space force one to use global optimization algorithms. Simulated annealing will be used to determine a good position for the changepoints.

If the joint density were separable then it would be the product of the marginal densities for each changepoint. Figure 5.11 shows the product of the marginal densities in this example. Comparing figure 5.11 with figure 5.10 shows that the joint density is only separable to a crude approximation. The disagreement between figure 5.11 and figure 5.10 is strongest along the axis of symmetry. This is hardly surprising, since the interaction between neighbouring changepoints is strongest if they are closely spaced.

As mentioned previously, the parameter space may be partitioned into three subspaces; linear coefficients, noise parameter(s) and changepoints. The Gibbs sampler provides an easily applied approach for sampling the parameter space. This sampling strategy involves sampling each changepoint

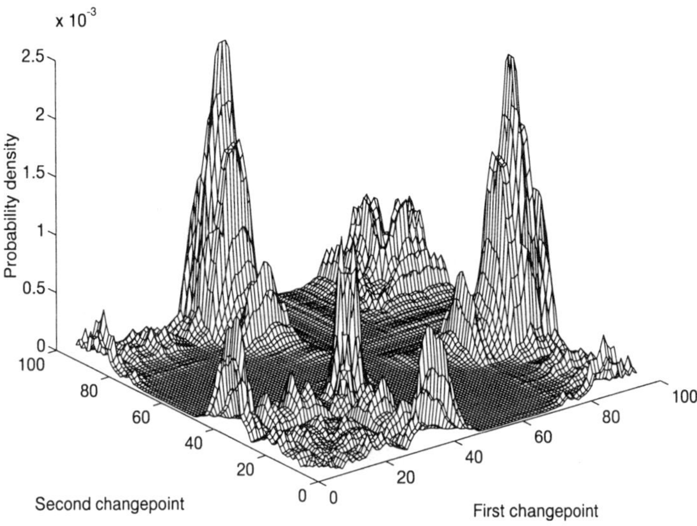

FIGURE 5.10. Joint posterior density for two changepoints

in turn rather than altogether. This can be quite an efficient approach because of the lack of posterior correlations between changepoints as exhibited by figure 5.10.

Note that, from the point of view of a Gibbs sampler, the joint pdf is always proportional to the conditional pdf for reasons that are explained in Appendix E.3. In almost all cases it will not be possible to sample the conditional densities directly. In such cases, a hybrid sampling strategy will be employed which rotates Gibbs sampling steps and local Metropolis steps in sequence. In the absence of the conditional density, the Gibbs sampler can sample either from some non-normalized density that is proportional to the conditional density. In fact the Gibbs sampler can be implemented by sampling each parameter in turn directly from the joint density.

Stephens [121, 122] uses the key property of the Gibbs sampler, which is the reduction of the effective dimensionality of a sampling problem, to great effect. He shows how a data sequence that exhibits multiple-changepoints can be analysed using a single changepoint model similar to that described in section 5.2. A changepoint problem can be broken down in this way because the pdf for the position of a changepoint is only directly dependent

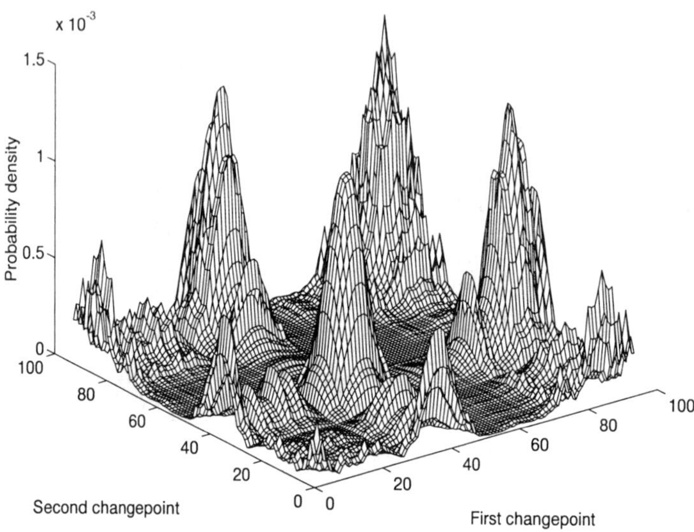

FIGURE 5.11. Product of two marginal densities

on the position of the changepoint immediately before and the one immediately after (except for a scaling factor that does not affect sampling). Sampling a conditional density to update one single changepoint at a time is obviously more straightforward than sampling from the joint density of all changepoints.

In some situations, such as in Gaussian noise environments, the dimensionality may be reduced by integrating out nuisance parameters. Using this approach implies using marginal estimates rather than joint estimates to infer the positions of the changepoints. The simplest case that will be studied is additive Gaussian noise where both linear parameters and noise parameters can be integrated out so the sample space of the Gibbs sampler is the space of changepoints. This is the situation in figure 5.10.

The applicability of the Gaussian assumption is governed primarily by the impulsiveness of the noise process corrupting the data. As noted by many authors [2, 75, 112], when noise corruption is impulsive it is often useful to assume a probability density $p(\mathbf{x}, \alpha) \; : \; \mathbf{x} \in \Re, \, \alpha > \mathbf{0}$ in the L_q

norm family of densities for which:

$$p(\mathbf{x} \mid \alpha) = \frac{\alpha^{\frac{N}{q}}}{2\frac{1}{q}\Gamma\left(\frac{1}{q}\right)} \exp\left(-\alpha \sum_{i=1}^{N} |x_i|^q\right) \tag{5.34}$$

Setting $q = 2$ gives a Gaussian density, and $q = 1$ gives a Laplacian density. The more impulsive is the noise the lower is the appropriate value of q.

In this case, it is possible to integrate out the noise parameter α. Assuming a Jeffreys' prior for the scale parameter α and integrating out α from 5.34 gives:

$$\begin{aligned} p(\mathbf{x}) &\propto \frac{\Gamma\left(\frac{N}{q}\right)}{2\frac{1}{q}\Gamma\left(\frac{1}{q}\right)} \left[\sum_{i=1}^{N} |x_i|^q\right]^{-\frac{N}{q}} \\ &\propto \left[\sum_{i=1}^{N} |x_i|^q\right]^{-\frac{N}{q}} \end{aligned}$$

This means that, for a given value of q, the Gibbs sampler need only sample the space of changepoints $\{m\}$ and linear coefficients $\mathbf{b}$.

In general the noise parameters may not be integrated out and therefore it is necessary to obtain samples of the noise parameters as well.

5.6.1 Sampling changepoint space

A changepoint m is restricted to be an integer in the range $\mathbf{S} = \{1 \leq m \leq N\}$. If there are M changepoints in the signal model then the changepoint space is given by the grid of discrete points in $\mathbf{S}^M$. All changepoints must be distinct.

The transition to a new state in the Metropolis algorithm may be implemented by drawing a uniform random integer in the range $\mathbf{S}$. The simplest way to do this is to generate a random uniform variate and scale it to cover the range and round off. This is quite adequate for small $N < 100$, but is a very inefficient means of sampling the changepoint space for large $N > 1000$. This is because of the random clustering effect [95] exhibited by uniform random variates.

Quasi-random sequences provide a far superior method for sampling changepoint space. Sobol' sequences [95] are used here. A Sobol' sequence is uniformly distributed over the range (0, 1] and on examination looks very much like a random sequence. The difference is that Sobol' sequences are self-avoiding, being cleverly constructed so new variates in the sequence fill gaps left behind by previous variates. As the number of samples increase then the space gets sampled in finer detail. The application to finding changepoints is obvious. If a position has been tested previously as a possible location for a changepoint then it is preferable to "try somewhere else" later, and as far away from other changepoints as possible.

5.6.2 *Sampling linear parameter space*

In many cases it is not possible to integrate out the linear parameters $\mathbf{b}$. This implies that some means of sampling the linear parameter space must be found because of the necessity of including it in the Gibbs sampler base transitions. In such cases one usually finds that it is not possible to sample from the conditional density $p(\mathbf{b} \mid \sigma, \{m\}, \mathbf{I})$ directly in which case the Metropolis algorithm may be required. As discussed in section 4.5 the Metropolis algorithm works best if the standard deviation of the proposal density closely matches that of the density being simulated.

Gaussian approximation

A method for constructing a proposal density will now be described. The proposal density is based on a Gaussian approximation to the pdf being sampled:

$$\log p(\mathbf{b} \mid \{m\}, \sigma, \mathbf{d}, \mathbf{I}) \approx -N\left(2\,\pi\,\sigma^2\right) - \frac{1}{2\,\sigma^2}(\mathbf{d} - \mathbf{Gb})^{\mathbf{T}}(\mathbf{d} - \mathbf{Gb}) \quad (5.35)$$

The Hessian matrix is given by:

$$\nabla\nabla \log p(\mathbf{b} \mid \{m\}, \sigma, \mathbf{d}, \mathbf{I}) = -\frac{\mathbf{G}^{\mathbf{T}}\mathbf{G}}{\sigma^2}$$

where differentiation is taken with respect to the linear parameters $\mathbf{b}$. Changing the sign of the Hessian matrix and inverting gives an estimate for the covariance matrix

$$\mathbf{C} = \sigma^{\mathbf{2}}\left(\mathbf{G}^{\mathbf{T}}\mathbf{G}\right)^{-\mathbf{1}}$$

Sometimes the variance may not be readily available in which case it may be estimated using expression 2.31:

$$\sigma^2 = \frac{1}{N - M}\left[\mathbf{d}^{\mathbf{T}}\mathbf{d} - \mathbf{d}^{\mathbf{T}}\mathbf{G}\left(\mathbf{G}^{\mathbf{T}}\mathbf{G}\right)^{-1}\mathbf{G}^{\mathbf{T}}\mathbf{d}\right]$$

where M is the length of vector $\mathbf{b}$. Note that this estimate for the covariance matrix depends only on the changepoint position. This is a major advantage in using the Gaussian approximation.

Neal [82] notes that, during annealing, the curvature of the proposal density should match the canonical distribution. This is easily achieved by scaling the covariance matrix:

$$\mathbf{C}_T = T\,\mathbf{C} = T\,\sigma^2\left(\mathbf{G}^{\mathbf{T}}\mathbf{G}\right)^{-1}$$

where T is the temperature.

This covariance matrix for the proposal density is derived under the assumption that the canonical distribution may be locally approximated by a Gaussian density, from which curvature information may be extracted. Ingber and Rosen [53] use curvature information in their very fast simulated reannealing algorithm, but determine curvature by some form of numerical differentiation.

The truth about whether or not the Gaussian approximation is a good approximation is not critical since the distribution produced by the Metropolis algorithm is independent of the proposal density[1]. The aim here is to keep the rejection rate as low as possible; the Gaussian approximation is a compromise between tractability and quality of approximation.

Colouring

Armed with the information about the covariance matrix, one is now in a position to produce candidate states using the following relation:

$$\mathbf{b}' = \mathbf{b} + \mathbf{C}_T^{1/2}\,\mathbf{p} \tag{5.36}$$

where $\mathbf{p}$ is a random vector whose elements are statistically independent, sampled from a density that is positive[2] everywhere in parameter space. The probability density of each element should be comparable to a zero mean unit variance Gaussian density. The origin of expression 5.36 is easily understood from the material in appendix B which describes how to draw random samples from a multivariate Gaussian distribution.

The "obvious" choice for the distribution for $\mathbf{p}$ is Gaussian. This would constitute "Boltzmann annealing". Szu and Hartley [125] propose using Cauchy variates for generating candidate states. This is the basis of their "fast annealing" algorithm. The advantage of using Cauchy variates is that occasional large transitions are possible. This can lead to considerable improvements in the convergence of the algorithm if the starting position is poor.

A more conservative approach to generating candidate states is to change only one component of $\mathbf{b}$ at a time. This is useful in more difficult problems where one cannot make significant gains by changing all components at once because the new state is almost always rejected. In this case the new state (generated by changing just the jth component of $\mathbf{b}$) may be proposed using:

$$b_j{}' = b_j + \sigma_j\, p$$

where σ_j is the square root of the diagonal element C_{jj} of the covariance matrix $\mathbf{C_T}$ above. This gives the standard deviation σ_j of the parameter b_j when all other linear coefficients $\{b_j \,:\, i \neq j\}$ are integrated out of the Gaussian approximation in expression 5.35.

[1]The proposal density is usually assumed to be constant. That condition has been relaxed here because the main objective is optimization rather than random simulation.

[2]This ensures that the Markov chain is irreducible.

5.6.3 Sampling noise parameter space

Sampling noise parameters from the joint density can be difficult. One approach that may be used is to estimate the noise parameters from the innovation sequence $\mathbf{e}$ where $\mathbf{e} = \mathbf{d} - \mathbf{G}\mathbf{b}$ as defined in equation 5.14. The next step would be to draw random samples from about the mode using information obtained from the Hessian matrix.

5.7 Multiple changepoint results

5.7.1 Synthetic step data

In this section, the performance of the multiple changepoint detector is examined using synthetic data. Consider the data plotted in figure 5.12 which shows a piecewise constant signal corrupted by Laplacian noise.

First, the data set was analysed using a Gibbs sampler assuming that the noise corrupting the data was *Laplace* distributed. The parameters were initialized to random values and one hundred iterations of the Gibbs

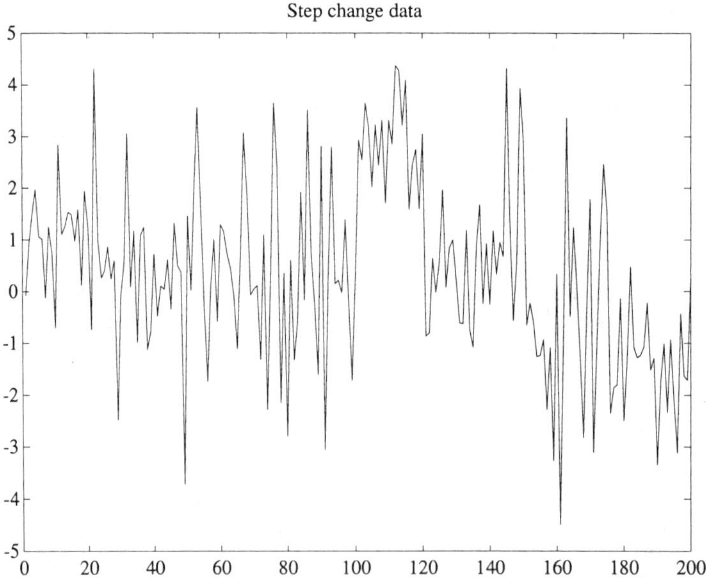

FIGURE 5.12. Step change data with Laplacian noise added

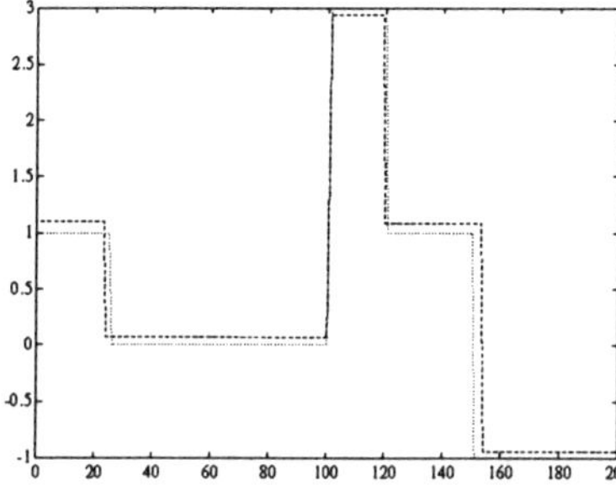

FIGURE 5.13. Laplacian noise assumed

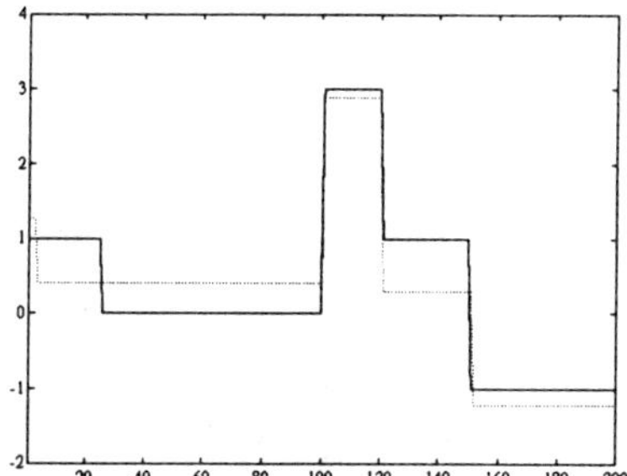

FIGURE 5.14. Gaussian noise assumed

sampler carried out. The reconstructed signal (dotted) is compared with the true signal (solid) in figure 5.13. The correspondence is quite good, particularly when one takes into consideration the number of data and the noise level.

Second, the data set was analysed using a Gibbs sampler assuming that the noise corrupting the data was *Gaussian* distributed. The parameters were initialized to the same values as in the Laplacian case and one hundred iterations of the Gibbs sampler carried out. The reconstructed signal (dotted) is compared with the true signal (solid) in figure 5.14. The results are noticeably poorer in this case.

5.7.2 Well log data

In this section, the performance of the changepoint detector is investigated using *well log data.* This geophysical data contains measurements of nuclear magnetic response and conveys information about rock structure and, in particular, the boundaries between different rock strata. The data set is plotted in figure 5.15.

The first thing of note upon examination of the data is that there is a considerable number of outliers. The noise corrupting the underlying step change signal is quite impulsive so a Gaussian noise assumption would seem to be unjustified. Ten changepoints are reasonably clear but there seem to be two or three smaller changepoints as well.

Explicit assumptions regarding the models describing the data are necessary in order to carry out a Bayesian analysis. Laplacian noise is assumed for the noise because of its impulsiveness. A thirteen changepoint model is also assumed to take into account the ten *obvious* changepoints and to leave the possibility open for three more.

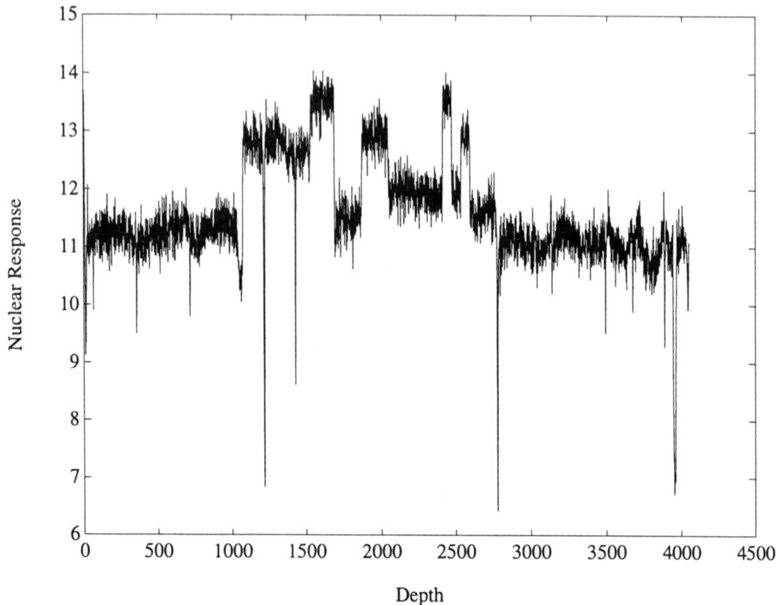

FIGURE 5.15. Well log data

The outliers were removed by eliminating the forty greatest outliers from the data. The (adjusted) data set is plotted in figure 5.16. The results of a Bayesian analysis involving two hundred iterations of a Gibbs sampler are also shown in figure 5.16. The correspondence between the two is quite good except for the "missed" changepoints on either side of the narrow peak at $m \approx 2500$.

The starting point for the Gibbs sampler in joint changepoint-linear parameter space was chosen at random. The progress of the Gibbs sampler in optimizing the joint posterior probability density is shown in figure 5.17. From the start point to the one hundredth iteration of the Gibbs sampler the probability density increases by a factor of 10^{50}. This result appears even better when one considers that the total dimensionality of the parameter space is 27 and that the function being optimized has a very large number of local minima. After about one hundred and fifty iterations the Gibbs sampler converged to its near optimal solution shown in figure 5.16.

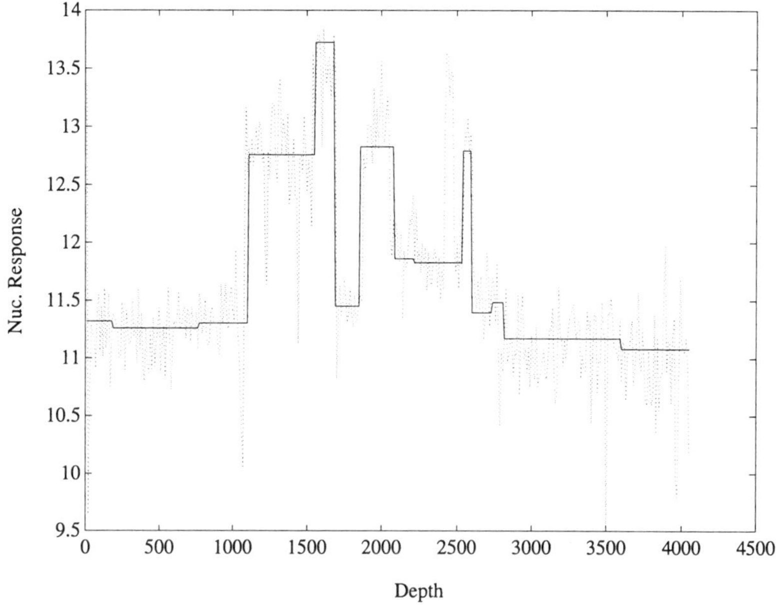

FIGURE 5.16. Well log data and inferred step changes

5.8 Concluding Remarks

In this chapter, it was shown how abrupt changes in the linear parameters of a signal may be modelled and how the positions of changepoints can be inferred using Bayesian methods. The data set is modelled as a signal consisting of a linear weighted sum of basis functions corrupted by added noise of known distribution. The detection of the location in time of abrupt changes in these linear coefficients was considered. The changepoint problems that were investigated include changes in the mean of a Gaussian process and changes in the coefficients of autoregressive models.

For Gaussian noise environments the posterior density of the changepoints can be derived in closed form. This means that their position in time can be inferred independently of the values of the linear parameters. Simulations involving step changes and autoregressive models have given excellent results.

A recursive procedure based on the Woodbury formula, has also been derived. The application of this procedure can result in very significant reductions in computation time.

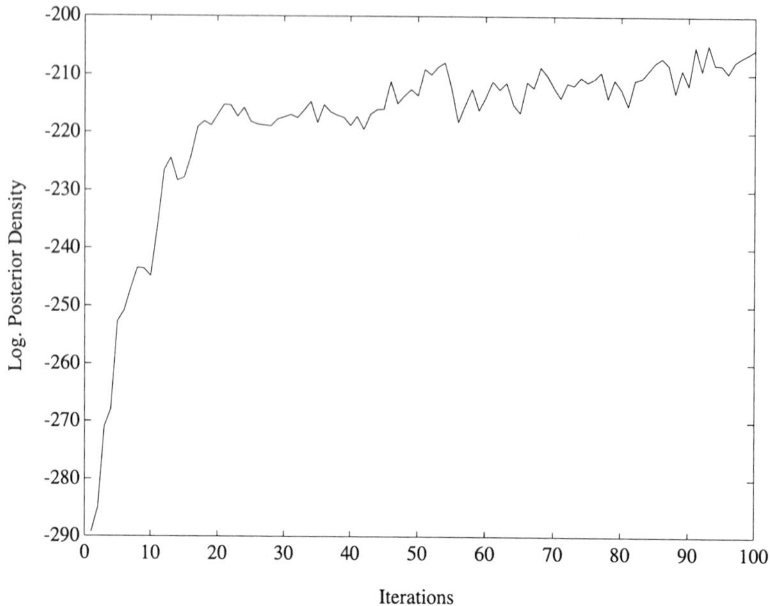

FIGURE 5.17. Simulated annealing – Probability versus number of iterations

The Bayesian approach exhibits some interesting features. The width of the peak of the posterior density indicates the degree of confidence in the detection. In those cases where a null detection is returned the peak tends to appear close to the endpoints of the time series.

Ó Ruanaidh and Fitzgerald [85, 86] have applied the changepoint detector to the location of changepoints in cubic spline data. The detection of changepoints in polynomial data is an ill-posed problem and, in experiments, it was found that a diffuse nature marginal density, obtained when the detector was not confident, was an invaluable aid for avoiding false detections.

The model was extended to accommodate non-Gaussian noise environments, determining the position of changepoints using simulated annealing. A hybrid sampling technique combining the Gibbs sampler and the local Metropolis-Hastings algorithm was used to sample the canonical distribution. The method of simulated annealing was successful in determining the main features in the data. For synthetic data the results were very good. For all the examples discussed, the "true" fit (i.e. the signal actually used to generate the data) was less probable than the fit found using simulated annealing. The results obtained for real data are more difficult to assess.

6

Restoration of Missing Samples in Digital Audio Signals

The aim of this chapter is to describe a novel method for interpolating autoregressive data. This is applied to the restoration of missing samples in digital audio signals. The section of audio signal in question is modelled as a stationary autoregressive process, and missing samples are imputed using the Gibbs sampler. The corresponding ML and EM algorithm solutions to the problem are developed and discussed, and the results are compared for both real and synthetic data.

6.1 Introduction

Clicks are a familiar problem in audio gramophone signals, and take the form of sudden unexpected bursts of impulsive noise with random but finite duration. These bursts of noise have numerous causes such as dirt, electrical interference or mechanical damage to the storage medium. The original signal is often effectively lost. Several methods of detecting clicks have been devised [40, 136], with the best approaches being model based. Once a click has been detected the "suspect" samples are removed and replaced by interpolation.

For this interpolation problem, we consider the data samples $\{x_i : 1 \leq i < m\}$ and $\{x_i : m+l \leq i \leq N\}$ as "observed" data, and the data samples $\{x_i : m \leq i < m+l\}$ as "missing" data (or "hidden" data). For conciseness, we partition the data into two vector components consisting of the missing data vector $\mathbf{z}$, and the observed data vector $\mathbf{y}$. The combined vector of all samples $\{x_i : 1 \leq i \leq n\}$ will be called the augmented data vector $\mathbf{x}$. The interpolation problem may now be stated as: given the observed data $\mathbf{y}$ infer the value of the missing data $\mathbf{z}$, or equivalently, use the observed components of $\mathbf{x}$ to infer its unknown missing components.

6.2 Model formulation

The autoregressive (AR) model has been found to be an effective means of modelling audio signals [40, 136, 137]. It is assumed that the samples of augmented data $\{x_i : p+1 \leq i \leq n\}$, which we denote as $\mathbf{w}$, satisfy an AR model of order p

$$x_i = \sum_{j=1}^{p} \theta_i \, x_{i-j} + e_i \tag{6.1}$$

and that the excitation sequence $\mathbf{e}$ is drawn from an independent identically distributed (i.i.d.) Gaussian random process with zero mean and standard deviation σ. We may write the excitation sequence $\mathbf{e}$ in the form

$$\mathbf{e} = \mathbf{w} - \mathbf{L}\,\theta \tag{6.2}$$

where θ is the vector of AR parameters and where $\mathbf{L} = \mathbf{L}(\mathbf{x})$ is an $(N - p) \times p$ matrix function of the augmented data. Alternatively, one may write the excitation sequence as

$$\mathbf{e} = \mathbf{K}\,\mathbf{x} \tag{6.3}$$

where $\mathbf{K} = \mathbf{K}(\theta)$ is an $N \times N$ matrix function of the AR parameters.

Let us assume for example that the AR model is fourth order which implies $p = 4$. Equation 6.2 can be expanded as follows:

$$\begin{bmatrix} e_5 \\ e_6 \\ e_7 \\ \vdots \\ e_{M-1} \\ e_M \\ e_{M+1} \\ \vdots \\ e_N \end{bmatrix} = \begin{bmatrix} x_5 \\ x_6 \\ x_7 \\ \vdots \\ x_{M-1} \\ x_M \\ x_{M+1} \\ \vdots \\ x_N \end{bmatrix} - \begin{bmatrix} x_4 & x_3 & x_2 & x_1 \\ x_5 & x_4 & x_3 & x_2 \\ x_6 & x_5 & x_4 & x_3 \\ \vdots & \vdots & \vdots & \vdots \\ x_{M-2} & x_{M-3} & x_{M-4} & x_{M-5} \\ x_{M-1} & x_{M-2} & x_{M-3} & x_{M-4} \\ x_M & x_{M-1} & x_{M-2} & x_{M-3} \\ \vdots & \vdots & \vdots & \vdots \\ x_{N-1} & x_{N-2} & x_{N-3} & x_{N-4} \end{bmatrix} \begin{bmatrix} \theta_1 \\ \theta_2 \\ \theta_3 \\ \theta_4 \end{bmatrix} \tag{6.4}$$

Note that the residuals are taken as beginning with e_{p+1} rather than e_1. This is because, in an AR model, each sample depends on p previous samples. Starting with e_{p+1} avoids the complication of making the residual dependent on unobserved data, which occurred before the data record begun. Therefore, it is arguably better in most cases to omit $e_1 \ldots e_p$ and to work only with observed data only. It is possible to expand the system by setting unobserved data to zero (i.e. the expected value). This is sometimes useful in some situations. This point will be discussed further in section 6.5.

If unobserved data is taken to be zero then equation 6.3 can be expanded as follows:

$$\begin{bmatrix} e_1 \\ e_2 \\ e_3 \\ e_4 \\ \vdots \\ e_{N-3} \\ e_{N-2} \\ e_{N-1} \\ e_N \end{bmatrix} = \begin{bmatrix} 1 & 0 & 0 & 0 & & 0 & 0 & 0 & 0 \\ -\theta_1 & 1 & 0 & 0 & & 0 & 0 & 0 & 0 \\ -\theta_2 & -\theta_1 & 1 & 0 & & 0 & 0 & 0 & 0 \\ -\theta_3 & -\theta_2 & -\theta_1 & 1 & & 0 & 0 & 0 & 0 \\ & & & & \ddots & & & & \\ 0 & 0 & 0 & 0 & & 1 & 0 & 0 & 0 \\ 0 & 0 & 0 & 0 & & -\theta_1 & 1 & 0 & 0 \\ 0 & 0 & 0 & 0 & & -\theta_2 & -\theta_1 & 1 & 0 \\ 0 & 0 & 0 & 0 & & -\theta_3 & -\theta_2 & -\theta_1 & 1 \end{bmatrix} \begin{bmatrix} x_1 \\ x_2 \\ x_3 \\ x_4 \\ \vdots \\ x_{N-3} \\ x_{N-2} \\ x_{N-1} \\ x_N \end{bmatrix} \tag{6.5}$$

where $\mathbf{K}$ is a band diagonal Toeplitz matrix.

Also note that the residuals can be expressed either as a linear function of the missing data, or linearly in terms of the AR parameters. This is the key to the elegance of iterative algorithms based on the AR model.

6.2.1 *The likelihood and the excitation energy*

The aim of interpolation is to estimate the missing data $\mathbf{z}$. All three approaches to interpolation (ML, EM algorithm and Gibbs sampler) depend on the properties of the likelihood function.

Let the likelihood of the augmented data be given by

$$p(\mathbf{e}) = \prod_{i=p+1}^{N} p(e_i) = p(\mathbf{w} \mid \theta, \sigma)$$

From the assumptions made about the excitation process we have

$$p(\mathbf{w} \mid \theta, \sigma) = \left(2\pi\sigma^2\right)^{-\frac{(N-p)}{2}} \exp\left(-\frac{\mathbf{e}^{\mathbf{T}}\mathbf{e}}{2\sigma^2}\right) \tag{6.6}$$

Substituting for $\mathbf{e}$ from equation 6.2 we can express the excitation energy as a quadratic form in θ

$$\mathbf{e}^{\mathbf{T}}\mathbf{e} = \mathbf{w}^{\mathbf{T}}\mathbf{w} - 2\mathbf{w}^{\mathbf{T}}\mathbf{L}\,\theta + \theta^{\mathbf{T}}\,\mathbf{L}^{\mathbf{T}}\,\mathbf{L}\theta \tag{6.7}$$

Similarly, substituting for $\mathbf{e}$ from equation 6.3 we can express the excitation energy as a quadratic form in $\mathbf{z}$

$$\begin{aligned}\mathbf{e}^{\mathbf{T}}\mathbf{e} &= \mathbf{w}^{\mathbf{T}}\mathbf{K}^{\mathbf{T}}\mathbf{K}\mathbf{w} \\ &= \mathbf{y}^{\mathbf{T}}\mathbf{A}\mathbf{y} + 2\mathbf{y}^{\mathbf{T}}\mathbf{B}\mathbf{z} + \mathbf{z}^{\mathbf{T}}\mathbf{D}\,\mathbf{z}\end{aligned} \tag{6.8}$$

where the matrices $\mathbf{A}$, $\mathbf{B}$ and $\mathbf{D}$ are disjointed partitions of $\mathbf{K}^{\mathbf{T}}\mathbf{K}$. Let

$$\mathbf{K}^{\mathbf{T}}\mathbf{K} = \begin{bmatrix} \mathbf{A}_{11} & \mathbf{B}_1 & \mathbf{A}_{12} \\ \mathbf{B}_1^{\mathbf{T}} & \mathbf{D} & \mathbf{B}_2^{\mathbf{T}} \\ \mathbf{A}_{21} & \mathbf{B}_2 & \mathbf{A}_{22} \end{bmatrix} \tag{6.9}$$

then $\mathbf{A}$ and $\mathbf{B}$ are given by

$$\mathbf{A} = \begin{bmatrix} \mathbf{A}_{11} & \mathbf{A}_{12} \\ \mathbf{A}_{21} & \mathbf{A}_{22} \end{bmatrix} \qquad \mathbf{B} = \begin{bmatrix} \mathbf{B}_1 \\ \mathbf{B}_2 \end{bmatrix} \tag{6.10}$$

and where $\mathbf{w}$ and $\mathbf{y}$ are expressed as follows:

$$\mathbf{w} = \begin{bmatrix} \mathbf{y}_1 \\ \mathbf{z} \\ \mathbf{y}_2 \end{bmatrix} \qquad \mathbf{y} = \begin{bmatrix} \mathbf{y}_1 \\ \mathbf{y}_2 \end{bmatrix} \tag{6.11}$$

$\mathbf{K}^{\mathbf{T}}\mathbf{K}$ is band diagonal and is, of course, symmetric. In addition, $\mathbf{K}^{\mathbf{T}}\mathbf{K}$ is Toeplitz after the pth row.

The matrix $\mathbf{D}$ is therefore band diagonal, and if the partition is such that $\mathbf{D}$ contains no elements of the first p rows of $\mathbf{K}^{\mathbf{T}}\mathbf{K}$ then it will be Toeplitz as well. Also $\mathbf{A}$ is band diagonal but is never Toeplitz because it contains the first p rows.

In addition $\mathbf{B}$ is highly sparse if the order of the AR model is much less than the number of missing sample data.

Note: *For the sake of convenience, in the rest of this chapter N will be taken to be the number of residuals used to calculate the likelihood rather than the actual number of data.*

6.2.2 Maximum likelihood

The unknown parameters may be divided into three groups. The missing data may be grouped together to form a vector parameter $\mathbf{z}$ and the autoregressive parameters to form a vector parameter θ. The remaining parameter is the standard deviation σ which is a scalar quantity.

Differentiating expression 6.6 with respect to σ we can determine the maximum likelihood estimate for σ as

$$\hat{\sigma} = \sqrt{\frac{\mathbf{e}^{\mathbf{T}}\mathbf{e}}{N}} \tag{6.12}$$

The missing data and the autoregressive parameters may also be obtained by maximizing the likelihood in expression 6.6. This is exactly equivalent to minimizing the excitation energy $\mathbf{e}^{\mathbf{T}}\,\mathbf{e}$.

If the missing data are fixed then one may maximize expression 6.7 with respect to θ to estimate the AR parameters as

$$\hat{\theta} = \left(\mathbf{L}^{\mathbf{T}}\,\mathbf{L}\right)^{-1}\mathbf{L}^{\mathbf{T}}\,\mathbf{w} \tag{6.13}$$

If the AR parameters are fixed then one may maximize expression 6.8 with respect to $\mathbf{z}$ to estimate the missing data as

$$\hat{\mathbf{z}} = -\mathbf{D}^{-1}\,\mathbf{B}^{\mathbf{T}}\,\mathbf{y} \tag{6.14}$$

Vaseghi [136] and Veldhuis [137] both describe a method for jointly estimating the AR parameters and the missing data by successively maximizing the likelihood with respect to θ, and then with respect to $\mathbf{z}$. The procedure is iterated until the values converge at a maximum of the likelihood function $p(\mathbf{w} \mid \theta,\, \sigma)$. This is obviously a multivariable version of the simple alternating variables method of optimization discussed in section 3.2.1.

6.3 The EM algorithm

In this section, we give a brief description of the Expectation Maximization (EM) algorithm [29], and its application to the interpolation of autoregressive data. The EM algorithm is discussed and applied to several simple examples by Therrien [128] and Tanner [126]. The aim is to choose the parameter of interest, in this case $\mathbf{z}$, in order to maximize its predictive density $p(\mathbf{z} \mid \mathbf{y})$. An alternative formulation of the EM algorithm is presented by Veldhuis [137].

We begin by writing Bayes' theorem in the following form:

$$p(\mathbf{z} \mid \mathbf{y}) \;\propto\; \frac{p(\mathbf{z},\, \mathbf{y} \mid \theta)}{p(\theta \mid \mathbf{y},\, \mathbf{z})} \tag{6.15}$$

where it is assumed that $p(\mathbf{y})$ is constant because $\mathbf{y}$ is fixed, and $p(\theta)$ is uniform because of our ignorance about the value of θ in the absence of data.

Taking the logs of both sides we have

$$\log p(\mathbf{z} \mid \mathbf{y}) = \log p(\mathbf{z},\, \mathbf{y} \mid \theta) - \log p(\theta \mid \mathbf{y},\, \mathbf{z}) + k \tag{6.16}$$

where k is an additive constant (which may be ignored). This expression is log likelihood minus log posterior density.

Multiplying both sides by $p(\theta \mid \mathbf{y}, \mathbf{z}^*)$ and taking expectations with respect to θ we obtain

$$\log p(\mathbf{z} \mid \mathbf{y}) = \int_\Theta \log p(\mathbf{z} \mid \mathbf{y}, \theta)\, p(\theta \mid \mathbf{y}, \mathbf{z}^*)\, d\theta - \int_\Theta \log p(\theta \mid \mathbf{y}, \mathbf{z})\, p(\theta \mid \mathbf{y}, \mathbf{z}^*)\, d\theta \tag{6.17}$$

where $\Theta = \Re^q$ is the space of the autoregressive parameters and q is the order of the AR model. The left hand side of equation 6.17 remains unchanged because $\log p(\mathbf{z} \mid \mathbf{y})$ is not a function of θ.

Dempster, Laird and Rubin [29] define the Q function and the H function as follows:

$$Q(\mathbf{z}, \mathbf{z}^*) = \int_\Theta \log p(\mathbf{z} \mid \mathbf{y}, \theta)\, p(\theta \mid \mathbf{y}, \mathbf{z}^*)\, d\theta \tag{6.18}$$

$$H(\mathbf{z}, \mathbf{z}^*) = \int_\Theta \log p(\theta \mid \mathbf{y}, \mathbf{z})\, p(\theta \mid \mathbf{y}, \mathbf{z}^*)\, d\theta \tag{6.19}$$

Substituting into equation 6.17 gives

$$\log p(\mathbf{z} \mid \mathbf{y}) = Q(\mathbf{z}, \mathbf{z}^*) - H(\mathbf{z}, \mathbf{z}^*) \tag{6.20}$$

Rao [102] shows that

$$H(\mathbf{z}, \mathbf{z}^*) - H(\mathbf{z}, \mathbf{z}) \leq 0 \tag{6.21}$$

Dempster, Laird and Rubin [29] present an iterative scheme whereby if $\mathbf{z}_i = \mathbf{z}^*$ is the present iterate, then the next iterate $\mathbf{z}_{i+1} = \mathbf{z}$ is chosen to maximize the Q function. This value for $\mathbf{z}$ then becomes the new value for $\mathbf{z}^*$. Because of the inequality in equation 6.21, each new value for $\mathbf{z}^*$ is guaranteed not to decrease the value of the predictive density. The procedure is iterated until convergence when the predictive density can be increased no more, at which point a supremum has been located.

For our problem we can write the Gaussian log likelihood as:

$$\log p(\mathbf{z}, \mathbf{y} \mid \theta) = -\frac{\mathbf{e}^{\mathbf{T}}\, \mathbf{e}}{2\,\sigma^2} - \frac{N}{2} \log\left(2\,\pi\,\sigma^2\right) \tag{6.22}$$

From equation 6.3 we have $\mathbf{e} = \mathbf{w} - \mathbf{L}\,\theta$.

Let $\mathbf{L}_{i+1} = \mathbf{L}_{i+1}(\mathbf{y}, \mathbf{z}_{i+1})$ be the matrix that results from using missing data $\mathbf{z}_{i+1}$ and $\mathbf{L}_i = \mathbf{L}_i(\mathbf{y}, \mathbf{z}_i)$ be the corresponding matrix that results from using missing data $\mathbf{z}_i$.

The procedure will be as follows:

Expectation: Evaluate the Q function as an integral in closed form as a function of $\mathbf{z}_{i+1}$ (because $\mathbf{z}_i$ is fixed from the previous iteration).

Maximization: Maximize the Q function with respect to $\mathbf{z}_{i+1}$. The maximizer $\hat{\mathbf{z}}$ is the new iterate.

We begin by noticing that the posterior density $p(\theta \mid \mathbf{y}, \mathbf{z}^*)$ is essentially proportional to the likelihood $p(\mathbf{z}^*, \mathbf{y} \mid \theta)$.

The exact relationship is:

$$p(\theta \mid \mathbf{y}, \mathbf{z}^*) = p(\mathbf{z}^*, \mathbf{y} \mid \theta) \frac{p(\theta)}{p(\mathbf{z}^*, \mathbf{y})} \tag{6.23}$$

The expectation step involves integrating over θ and the maximization step involves differentiating with respect to $\mathbf{z}$. Therefore, only functions involving θ or $\mathbf{z}$ can affect the final result. The only term (other than the likelihood) on the right hand side that is a function of either θ or $\mathbf{z}$ is $p(\theta)$, and that term, as we stated earlier, is constant.

Noting that $p(\mathbf{z}_{i+1}, \mathbf{y} \mid \theta) = p(\mathbf{z}_{i+1} \mid \mathbf{y}, \theta) p(\mathbf{y} \mid \theta)$ we may write:

$$Q(\mathbf{z}_{i+1}, \mathbf{z_i}) \quad = \quad \int_\Theta \log p(\mathbf{z}_{i+1}, \mathbf{y} \mid \theta)\, p(\mathbf{z_i}, \mathbf{y} \mid \theta)\, d\theta \tag{6.24}$$

which is proportional to the original Q function in equation 6.18.

Now the log probability in equation 6.24 is defined as

$$\log p(\mathbf{z}_{i+1}, \mathbf{y} \mid \theta) \propto -\frac{(\mathbf{w} - \mathbf{L}_{i+1}\,\theta)^{\mathbf{T}} (\mathbf{w} - \mathbf{L}_{i+1}\,\theta)}{2\,\sigma^2} - \frac{N}{2} \log\left(2\,\pi\,\sigma^2\right) \tag{6.25}$$

where $\mathbf{w} = [\mathbf{y}_1, \mathbf{z}_{i+1}, \mathbf{y}_2]$.

The second term in equation 6.24 is defined as

$$p(\mathbf{z}_i, \mathbf{y} \mid \theta) \propto \left(2\,\pi\,\sigma^2\right)^{-\frac{N}{2}} \exp\left[-\frac{(\mathbf{v} - \mathbf{L}_i\,\theta)^{\mathbf{T}} (\mathbf{v} - \mathbf{L}_i\,\theta)}{2\,\sigma^2}\right] \tag{6.26}$$

where $\mathbf{v} = [\mathbf{y}_1, \mathbf{z}_i, \mathbf{y}_2]$.

It is required to integrate the product of expressions 6.25 and 6.26 over the autoregressive parameters θ, and maximize the integral with respect to $\mathbf{z}_{i+1}$.

6.3.1 Expectation

For convenience, we will denote $\mathbf{L}_{i+1}$ as $\mathbf{L}$ and $\mathbf{L}_i$ as $\mathbf{M}$ in order to clarify the algebra.

In appendix D we show that the integral for the Q function may be evaluated in closed form to give the following expression:

$$Q(\mathbf{z}_{i+1}, \mathbf{z_i}) = K(\mathbf{z}_i) \left[\frac{\mathrm{Trace}\left(\left(\mathbf{M}^{\mathbf{T}}\mathbf{M}\right)^{-1}\left(\mathbf{L}^{\mathbf{T}}\mathbf{L}\right)\right)}{2} \right.$$

$$+\frac{\left(\mathbf{w}-\mathbf{L}\,\hat{\phi}\right)^{\mathbf{T}}\left(\mathbf{w}-\mathbf{L}\,\hat{\phi}\right)}{2\,\sigma^2}-\frac{N}{2}\log\left(2\,\pi\,\sigma^2\right)\Bigg]$$

where

$$K\left(\mathbf{z}_i\right)=-\frac{\left(2\,\pi\,\sigma^2\right)^{-\frac{N}{2}}}{\sqrt{\det\mathbf{M}^{\mathbf{T}}\,\mathbf{M}}}\exp\left[-\frac{\left(\mathbf{v}^{\mathbf{T}}\,\mathbf{v}-\mathbf{v}^{\mathbf{T}}\,\mathbf{M}\left(\mathbf{M}^{\mathbf{T}}\,\mathbf{M}\right)^{-1}\mathbf{M}^{\mathbf{T}}\,\mathbf{v}\right)}{2\,\sigma^2}\right] \tag{6.27}$$

and

$$\hat{\phi}=\left(\mathbf{M}^{\mathbf{T}}\,\mathbf{M}\right)^{-1}\mathbf{M}^{\mathbf{T}}\,\mathbf{v} \tag{6.28}$$

6.3.2 Maximization

The next stage is to determine $\mathbf{z}_{i+1}$ such that $Q(\mathbf{z}_{i+1},\mathbf{z}_{\mathbf{i}})$ is maximized:

$$\frac{\partial}{\partial\mathbf{z}_{i+1}}\,Q(\mathbf{z}_{i+1},\mathbf{z}_{\mathbf{i}})=0 \tag{6.29}$$

This is equivalent to solving

$$\tfrac{1}{2}\,\mathrm{Trace}\left[\left(\mathbf{L_i}^{\mathbf{T}}\,\mathbf{L_i}\right)^{-1}\tfrac{\partial}{\partial\mathbf{z}_{i+1}}\left(\mathbf{L_{i+1}}^{\mathbf{T}}\,\mathbf{L_{i+1}}\right)\right]$$
$$+\tfrac{1}{2\,\sigma^2}\tfrac{\partial}{\partial\mathbf{z}_{i+1}}\left(\mathbf{w}-\mathbf{L}_{i+1}\,\hat{\phi}\right)^{\mathbf{T}}\left(\mathbf{w}-\mathbf{L}_{i+1}\,\hat{\phi}\right)=0 \tag{6.30}$$

We have already determined the second term on the left hand side of equation 6.30 in section 6.2.1. From equation 6.2 and equation 6.3 we can write

$$\mathbf{e}=\mathbf{w}-\mathbf{L}_{i+1}\,\hat{\phi}=\mathbf{K}\left(\hat{\phi}\right)\mathbf{w} \tag{6.31}$$

Then following the exact same line of argument as in section 6.2.1 we can show

$$\frac{1}{2\,\sigma^2}\frac{\partial}{\partial\mathbf{z}_{i+1}}\left(\mathbf{w}-\mathbf{L}_{i+1}\,\hat{\phi}\right)^{\mathbf{T}}\left(\mathbf{w}-\mathbf{L}_{i+1}\,\hat{\phi}\right)=\frac{1}{\sigma^2}\left(\mathbf{B}^{\mathbf{T}}\,\mathbf{y}+\mathbf{D}\,\mathbf{z_{i+1}}\right) \tag{6.32}$$

Differentiating the trace is no less straightforward. In appendix D.2 we show

$$\frac{1}{2}\,\mathrm{Trace}\left[\left(\mathbf{L_i}^{\mathbf{T}}\,\mathbf{L_i}\right)^{-1}\frac{\partial}{\partial\mathbf{z}_{i+1}}\left(\mathbf{L_{i+1}}^{\mathbf{T}}\,\mathbf{L_{i+1}}\right)\right]=\mathbf{T}\,\mathbf{z}_{i+1}+\mathbf{q} \tag{6.33}$$

where $\mathbf{T}$ is a symmetric band diagonal Toeplitz matrix. The diagonal elements of $\mathbf{T}$ equal the sum of the corresponding diagonal of $\left(\mathbf{L_i}^{\mathbf{T}}\,\mathbf{L_i}\right)^{-1}$.

The term in $\mathbf{q}$ depends on the observed data and may be calculated efficiently by means of a convolution.

Hence, the M step of the EM algorithm involves solving

$$\mathbf{T}\,\mathbf{z}_{i+1} + \mathbf{q} + \frac{1}{\sigma^2}\left(\mathbf{B}^{\mathbf{T}}\,\mathbf{y} + \mathbf{D}\,\mathbf{z}_{\mathbf{i+1}}\right) = 0 \tag{6.34}$$

In other words, we have to solve

$$\left(\sigma^2\,\mathbf{T} + \mathbf{D}\right)\mathbf{z}_{\mathbf{i+1}} = -\left(\sigma^2\,\mathbf{q} + \mathbf{B}^{\mathbf{T}}\,\mathbf{y}\right) \tag{6.35}$$

This is a band diagonal Toeplitz linear system of equations which can be solved in exactly the same way for the missing data stage in the ML system of equations in section 6.2.2.

6.4 Gibbs sampling

6.4.1 Description

The Gibbs sampler that will be developed here is quite unusual. Normally, Gibbs sampling is carried out by sampling one scalar parameter at a time. In this case, wherever convenient, vectors will be sampled rather than scalars. This approach to sampling is also advocated by Tanner [126].

The parameter space is partitioned into missing data $\mathbf{z}$, AR parameters θ and standard deviation σ as in section 6.2.1. The Gibbs sampler will be used to draw random variates from the joint density $p\,(\mathbf{z},\ \theta,\ \sigma \mid \mathbf{y})$.

Let us assume a starting position $(\mathbf{z}^0, \theta^0, \sigma^0)$ in parameter space. Consider

$$\begin{array}{lcl} \mathbf{z}^1 & \leftarrow & p\,(\mathbf{z} \mid \theta^0, \sigma^0, \mathbf{y}) \\ \theta^1 & \leftarrow & p\,(\theta \mid \sigma^0, \mathbf{z}^1, \mathbf{y}) \\ \sigma^1 & \leftarrow & p\,(\sigma \mid \mathbf{z}^1, \theta^1, \mathbf{y}) \\ \mathbf{z}^2 & \leftarrow & p\,(\mathbf{z} \mid \theta^1, \sigma^1, \mathbf{y}) \\ \theta^2 & \leftarrow & p\,(\theta \mid \sigma^1, \mathbf{z}^2, \mathbf{y}) \\ \vdots & & \vdots \\ \sigma^i & \leftarrow & p\,(\sigma \mid \mathbf{z}^i, \theta^i, \mathbf{y}) \end{array} \tag{6.36}$$

where the notation $\mathbf{z}^i \leftarrow p\,(\mathbf{z} \mid \theta^{i-1}, \sigma^{i-1}, \mathbf{y})$ means that a random sample $\mathbf{z}^i$ is drawn from the conditional probability density function of $\mathbf{z}$ with all other parameters fixed at the given values. It can be shown [36, 82] that the distribution of the random variates $(\mathbf{z}^i, \theta^i, \sigma^i)$ in parameter space asymptotically approaches the joint density $p\,(\mathbf{z},\ \theta,\ \sigma \,|\, \mathbf{y})$.

The approach described here is to generate random samples of missing data, autoregressive parameters and standard deviation individually and group them together in a single vector. In fact, from the point of view of audio restoration only the missing data component will be used. The autoregressive parameters and the standard deviation corresponding to the missing data are completely ignored. Because the three variables are drawn from the joint density, then as noted by Tanner [126] one variable taken on its own (a partition) is actually a sample from the marginal density. For this reason, the missing data is being drawn from the predictive density $p(\mathbf{z} \mid \mathbf{y})$.

6.4.2 Derivation of conditional densities

The conditional densities required for implementing the Gibbs sampler may be derived from the likelihood. If, for convenience, we assume uniform prior probabilities on σ, θ and $\mathbf{z}$ then we can write:

$$p(\mathbf{z},\, \theta,\, \sigma \mid \mathbf{y}) \propto p(\mathbf{z},\, \mathbf{y} \mid \theta,\, \sigma) \tag{6.37}$$

The conditional densities may be written as follows:

- In the case of the missing data we find that

$$p(\mathbf{z} \mid \theta,\, \sigma\, \mathbf{y}) = \frac{p(\mathbf{z},\, \theta,\, \sigma \mid \mathbf{y})}{\int_{\mathbf{Z}} p(\mathbf{z},\, \theta,\, \sigma \mid \mathbf{y})\, d\mathbf{z}} \tag{6.38}$$

  where $\mathbf{Z} = \Re^M$ where M is the number of data.

- For the autoregressive parameters we have

$$p(\theta \mid \sigma\, \mathbf{z}\, \mathbf{y}) = \frac{p(\mathbf{z},\, \theta,\, \sigma \mid \mathbf{y})}{\int_{\Theta} p(\mathbf{z},\, \theta,\, \sigma \mid \mathbf{y})\, d\theta} \tag{6.39}$$

  where $\theta = \Re^p$ where p is the order of the AR model.

- Therefore, the conditional density for the standard deviation is:

$$p(\sigma \mid \mathbf{z},\, \theta,\, \mathbf{y}) = \frac{p(\mathbf{z},\, \theta,\, \sigma \mid \mathbf{y})}{\int_{\mathbf{\Sigma}} p(\mathbf{z},\, \theta,\, \sigma \mid \mathbf{y})\, d\sigma} \tag{6.40}$$

  where $\mathbf{\Sigma} = (0,\, \infty)$.

It is not strictly necessary to compute the denominators in the equations above in order to sample from the conditional densities. The denominators are simple scaling factors and do not affect random sampling. In the following sections the conditional densities will be calculated in full because the resultant expressions are quite elegant and will serve to elucidate certain aspects of sampling from the conditional densities.

6.4.3 Conditional density for the missing data

From equation 6.8 we have:

$$p(\mathbf{z},\,\theta,\,\sigma\mid\mathbf{y}) \propto \sigma^{-N}\exp\left[-\frac{1}{2\,\sigma^2}\left(\mathbf{y}^{\mathbf{T}}\mathbf{A}\mathbf{y}+2\mathbf{y}^{\mathbf{T}}\mathbf{B}\mathbf{z}+\mathbf{z}^{\mathbf{T}}\,\mathbf{D}\,\mathbf{z}\right)\right] \tag{6.41}$$

Integrating out the missing data gives:

$$p(\theta,\,\sigma\mid\mathbf{y}) \propto \sigma^{-(N-M)}\exp\left[-\frac{1}{2\,\sigma^2}\left(\mathbf{y}^{\mathbf{T}}\mathbf{A}\mathbf{y}-\mathbf{y}^{\mathbf{T}}\,\mathbf{B}\,\mathbf{D}^{-1}\,\mathbf{B}^{\mathbf{T}}\,\mathbf{y}\right)\right] \tag{6.42}$$

where M is the number of missing data. Therefore:

$$p(\mathbf{z}\mid\theta,\,\sigma,\,\mathbf{y}) \propto \sigma^{-M}\exp\left[-\frac{Q_{\mathbf{z}}}{2\,\sigma^2}\right] \tag{6.43}$$

where

$$Q_{\mathbf{z}} = \mathbf{z}^{\mathbf{T}}\,\mathbf{D}\,\mathbf{z}+2\,\mathbf{y}^{\mathbf{T}}\mathbf{B}\mathbf{z}+\mathbf{y}^{\mathbf{T}}\,\mathbf{B}\,\mathbf{D}^{-1}\,\mathbf{B}^{\mathbf{T}}\,\mathbf{y} \tag{6.44}$$

We can write the conditional density $p(\mathbf{z}\mid\theta,\,\sigma,\,\mathbf{y})$ in the form

$$p(\mathbf{z}\mid\theta,\,\sigma,\,\mathbf{y}) \propto \exp\left[-\frac{1}{2\,\sigma^2}\,(\mathbf{z}-\hat{\mathbf{z}})^{\mathbf{T}}\,\mathbf{C}^{-1}\,(\mathbf{z}-\hat{\mathbf{z}})\right] \tag{6.45}$$

The conditional density is multivariate Gaussian with the inverse covariance matrix given by:

$$\mathbf{C}^{-1} = \frac{\mathbf{D}}{\sigma^{\mathbf{2}}} \tag{6.46}$$

The mode $\hat{\mathbf{z}}$ of the conditional density in expression 6.43 is exactly the same as that obtained in equation 6.14 when we were discussing the ML procedure.

$$\hat{\mathbf{z}} = -\mathbf{D}^{-1}\,\mathbf{B}^{\mathbf{T}}\,\mathbf{y} \tag{6.47}$$

Let $\mathbf{S}$ be the Cholesky decomposition of the inverse covariance matrix $\mathbf{C}^{-\mathbf{1}}$. Since $\mathbf{D}$ is band diagonal then so too is the inverse covariance matrix.

6.4.4 Conditional density for the autoregressive parameters

The conditional density for the autoregressive parameters works out in a similar way as for the missing data.

From equation 6.7 we have:

$$p(\mathbf{z},\,\theta,\,\sigma\mid\mathbf{y}) \propto \sigma^{-N}\exp\left[-\frac{\left(\mathbf{w}^{\mathbf{T}}\mathbf{w}-2\mathbf{w}^{\mathbf{T}}\mathbf{L}\,\theta+\theta^{\mathbf{T}}\,\mathbf{L}^{\mathbf{T}}\,\mathbf{L}\,\theta\right)}{2\,\sigma^2}\right] \tag{6.48}$$

Integrating out the autoregressive parameters gives:

$$p(\mathbf{z}, \sigma \mid \mathbf{y}) \propto \sigma^{-(N-p)} \exp\left[-\frac{\left(\mathbf{w}^{\mathbf{T}}\mathbf{w} - \mathbf{w}^{\mathbf{T}}\mathbf{L}\left(\mathbf{L}^{\mathbf{T}}\mathbf{L}\right)^{-1}\mathbf{L}^{\mathbf{T}}\mathbf{w}\right)}{2\,\sigma^2}\right] \tag{6.49}$$

where p is the number of autoregressive parameters. Therefore:

$$p(\theta \mid \mathbf{z}, \sigma, \mathbf{y}) \propto \sigma^{-p} \exp\left[-\frac{Q_\theta}{2\,\sigma^2}\right] \tag{6.50}$$

where

$$Q_\theta = \mathbf{w}^{\mathbf{T}}\mathbf{L}\left(\mathbf{L}^{\mathbf{T}}\mathbf{L}\right)^{-1}\mathbf{L}^{\mathbf{T}}\mathbf{w} - 2\mathbf{w}^{\mathbf{T}}\mathbf{L}\,\theta + \theta^{\mathbf{T}}\mathbf{L}^{\mathbf{T}}\mathbf{L}\,\theta \tag{6.51}$$

It is readily seen that the conditional density is multivariate Gaussian with inverse covariance matrix

$$\mathbf{C}^{-1} = \frac{\mathbf{L}^{\mathbf{T}}\mathbf{L}}{\sigma^2} \tag{6.52}$$

The mode $\hat{\theta}$ of the conditional density in expression 6.50 is exactly the same as that obtained in equation 6.13 when we were discussing the ML procedure:

$$\hat{\theta} = \left(\mathbf{L}^{\mathbf{T}}\mathbf{L}\right)^{-1}\mathbf{L}^{\mathbf{T}}\mathbf{w} \tag{6.53}$$

6.4.5 Conditional density for the standard deviation

The conditional density for the standard deviation presents a slightly more complicated situation than either for the missing data or the autoregressive parameters because it is a non-Gaussian density.

$$p(\sigma, \theta, \mathbf{z} \mid \mathbf{y}) = \sigma^{-N} \exp\left[-\frac{\mathbf{e}^{\mathbf{T}}\mathbf{e}}{2\,\sigma^2}\right] \tag{6.54}$$

We can integrate out the standard deviation to give

$$p(\theta, \mathbf{z} \mid \mathbf{y}) \propto \left[\mathbf{e}^{\mathbf{T}}\mathbf{e}\right]^{-\frac{N}{2}} \tag{6.55}$$

Combining the two terms gives

$$p(\sigma \mid \theta, \mathbf{z}, \mathbf{y}) \propto \left[\mathbf{e}^{\mathbf{T}}\mathbf{e}\right]^{\frac{N}{2}} \sigma^{-N} \exp\left[-\frac{\mathbf{e}^{\mathbf{T}}\mathbf{e}}{2\,\sigma^2}\right] \tag{6.56}$$

The term in equation 6.55 is in the form of a student-t distribution. The result is not a function of σ and is simply a scaling factor that can be ignored when sampling the conditional density in equation 6.56.

6.5 Implementation issues

In this section we describe some of the methods used to implement the ML, EM and Gibbs algorithms efficiently. We begin by discussing the estimation of AR parameters.

6.5.1 *Estimating AR parameters*

Consider the following alternative formulation of equation 6.4:

$$\begin{bmatrix} e_1 \\ e_2 \\ e_3 \\ e_4 \\ \vdots \\ e_{M-1} \\ e_M \\ e_{M+1} \\ \vdots \\ e_N \\ e_{N+1} \\ e_{N+2} \\ e_{N+3} \\ e_{N+4} \end{bmatrix} = \begin{bmatrix} x_1 \\ x_2 \\ x_3 \\ x_4 \\ \vdots \\ x_{M-1} \\ x_M \\ x_{M+1} \\ \vdots \\ x_N \\ 0 \\ 0 \\ 0 \\ 0 \end{bmatrix} - \begin{bmatrix} 0 & 0 & 0 & 0 \\ x_1 & 0 & 0 & 0 \\ x_2 & x_1 & 0 & 0 \\ x_3 & x_2 & x_1 & 0 \\ \vdots & \vdots & \vdots & \vdots \\ x_{M-2} & x_{M-3} & x_{M-4} & x_{M-5} \\ x_{M-1} & x_{M-2} & x_{M-3} & x_{M-4} \\ x_M & x_{M-1} & x_{M-2} & x_{M-3} \\ \vdots & \vdots & \vdots & \vdots \\ x_{N-1} & x_{N-2} & x_{N-3} & x_{N-4} \\ x_N & x_{N-1} & x_{N-2} & x_{N-3} \\ 0 & x_N & x_{N-1} & x_{N-2} \\ 0 & 0 & x_N & x_{N-1} \\ 0 & 0 & 0 & x_N \end{bmatrix} \begin{bmatrix} \theta_1 \\ \theta_2 \\ \theta_3 \\ \theta_4 \end{bmatrix} \tag{6.57}$$

In this case, there are $N+p$ residuals where N is the length of recording and p is the order of the AR model. Samples of unobserved data are replaced with their expected values, namely zero.

The AR parameters are estimated in the usual way using:

$$\mathbf{L}^{\mathbf{T}}\mathbf{L}\hat{\theta} = \mathbf{L}^{\mathbf{T}}\mathbf{x} \tag{6.58}$$

where $\mathbf{x}$ is the full augmented data vector.

If $\mathbf{L}$ is written in this form then, as noted by Therrien [128], $\mathbf{L}^{\mathbf{T}}\mathbf{L}$ will be Toeplitz and positive definite. This has some interesting implications. It is clear that in order to implement the Gibbs sampler it is imperative that the covariance matrix be positive definite. This is satisfied by the model formulation in equation 6.57. Some formulations such as expression 6.4 of the AR model lead to the inverse covariance only being positive semi-definite. This point is discussed in more detail by Therrien [128]. In addition, the system of equations resulting from the use of equation 6.57 is Toeplitz which implies that the Levinson algorithm may be used.

For our application a certain amount of healthy pragmatism needs to be exercised when estimating AR parameters. The requirement that the covariance matrix be positive definite and the need for numerical efficiency means that in many ways using the formulation in equation 6.57 is superior to that in equation 6.4.

6.5.2 Implementing the ML algorithm

There are two stages to the ML algorithm. The first stage is to solve for the missing data using equation 6.14 and the second stage is to estimate the AR parameters using equation 6.13.

Missing data in ML algorithm

As it stands, implementing the first stage directly from equation 6.14 is a poor way to compute the missing data. The reason for this is that the number of elements in the inverse matrix $\mathbf{D}^{-1}$ equals the square of the number M of missing data. For very large gaps a large amount of memory is required to store the elements of $\mathbf{D}^{-1}$ which can exceed the memory capabilities of many machines. This situation may be avoided by taking account of the fact that $\mathbf{K}^{\mathbf{T}}\mathbf{K}$ in equation 6.8 is band diagonal. A simple scheme for the efficient storage of band diagonal matrices is presented by Golub and Van Loan [43] and again by Press *et al* [95].

More seriously, computing the inverse is very computationally intensive so is generally to be avoided. A far superior approach is to solve

$$\mathbf{D}\,\mathbf{z} = -\mathbf{B}^{\mathbf{T}}\,\mathbf{y} \tag{6.59}$$

The matrix $\mathbf{D}$ is band diagonal and symmetric with a bandwidth equal to the order p of the autoregressive model.

Providing the position of the start of the gap is greater than the order of the AR model, then as noted in section 6.2.1, the matrix $\mathbf{D}$ will be Toeplitz. Only one row of a Toeplitz matrix needs to be stored so this presents great savings in computer memory.

Golub and Van Loan [43] present two efficient approaches for the solution of both band diagonal and Toeplitz systems of equations:

Band LU decomposition exploits the band diagonal nature of the system. The number of computations required [43] is approximately $2\,M\,(p^2 + 2\,p)$.

Levinson's algorithm makes use of the fact that the matrix $\mathbf{D}$ is usually Toeplitz. The number of computations required is approximately $4\,M^2$.

Whichever of the two techniques is more efficient depends on the length of the AR model relative to the number of missing data. If $2\,p^2 > M$ approximately then Levinson's algorithm will be the more efficient of the two.

The right hand side of equation 6.59 also deserves careful consideration. The matrix $\mathbf{B}$ is highly sparse consisting almost wholly of zero elements. Non-zero elements of $\mathbf{B}$ all lie in a band. The width of this band is about twice the order of the AR model. The most efficient way to compute the left hand side is to multiply only the non-zero rows of $\mathbf{B}$ with the corresponding rows of $\mathbf{y}$ to solve equation 6.53. In addition, the elements of $\mathbf{B}$ have already been computed in $\mathbf{D}$.

AR parameters in ML algorithm

The second stage of the ML algorithm is to estimate the AR parameters using equation 6.13.

If $\mathbf{L}$ is written in the form of equation 6.57 then the covariance matrix will be symmetric and Toeplitz which means that the Levinson algorithm may be used. This leads to large savings in computation which is most noticeable when the order of the AR model is large.

6.5.3 Implementing the EM algorithm

The EM algorithm, except for the presence of an additional term, is exactly the same as the ML algorithm. The requirements are exactly the same, that is, to use numerical algorithms that are efficient both in terms of use of computer memory and in terms of numerical computation.

The only term that differentiates the EM algorithm from the ML algorithm is that one must solve equation 6.35 rather than equation 6.14. The terms in $\sigma^2\,\mathbf{T}$ and $\sigma^2\,\mathbf{q}$ have to be determined in order to implement the algorithm. At first sight it would appear difficult because of the need to compute the variance of the excitation process σ^2. However, this is not so since all that has to be done is to form $\mathbf{T}$ and $\mathbf{q}$ from $\sigma^2\left(\mathbf{L}_i^{\mathbf{T}}\,\mathbf{L}_i\right)^{-1}$ rather than from $\left(\mathbf{L}_i^{\mathbf{T}}\,\mathbf{L}_i\right)^{-1}$. The former is simply the estimated covariance matrix for the AR parameters and is readily available from "canned" routines.

6.5.4 Implementation of Gibbs sampler

Generation of missing data

In this subsection, we describe the steps required to sample missing data from the conditional density $p\left(\mathbf{z} \mid \theta, \sigma, \mathbf{y}\right)$ as required by the Gibbs sampler.

Let $\mathbf{S}$ be the Cholesky decomposition of the inverse covariance matrix $\mathbf{C}^{-1}$ in equation 6.46. Golub and Van Loan [43] give a routine for determining the Cholesky decomposition of a band diagonal matrix such as $\mathbf{D}$ above. Appendix B describes how to draw random vector samples from a multivariate Gaussian distribution.

The stages in generating a random sample vector of missing data $\mathbf{z}$ are therefore:

- Determine a square root $\mathbf{S}$ of the inverse covariance matrix $\mathbf{C}^{-1}$. The inverse covariance matrix is band diagonal and therefore a Gaxpy-Cholesky [43] decomposition can be used.

- Generate a Gaussian random vector $\mathbf{u}$ of the same length as the parameter vector $\mathbf{z}$. Each component should have unit variance and and be statistically independent of the other components.

- Solve $\mathbf{S}^{\mathbf{T}}\,\mathbf{n} = \mathbf{u}$ for $\mathbf{n}$. The matrix $\mathbf{S}$ is not Toeplitz so a band LU decomposition must be used.

- Solve $\mathbf{D}\,\hat{\mathbf{z}} = -\mathbf{B}^{\mathbf{T}}\,\mathbf{y}$ as described in section 6.2.2.

- Compute $\mathbf{z} = \hat{\mathbf{z}} + \mathbf{n}$.

Generation of autoregressive parameters

The conditional density $p\,(\theta \mid \mathbf{z},\, \sigma,\, \mathbf{y})$ for the AR parameters θ is also multivariate Gaussian with inverse covariance matrix given in equation 6.52.

Using the methods described in appendix B leads to the following steps for generating a random sample vector of autoregressive parameters θ:

- Determine the square root $\mathbf{S}$ of the inverse covariance matrix $\mathbf{C}^{-1}$. This can be carried out using a standard Cholesky decomposition providing the covariance matrix is *positive definite.*

- Generate a Gaussian random vector $\mathbf{u}$ with each component of unit variance and with the same length as the parameter vector θ.

- Solve $\mathbf{S}^{\mathbf{T}}\,\mathbf{n} = \mathbf{u}$ for $\mathbf{n}$.

- Solve $\left(\mathbf{L}_i^{\mathbf{T}}\,\mathbf{L}_i\right)\hat{\theta} = \mathbf{L}_i^{\mathbf{T}}\,\mathbf{w}$ as described in section 6.2.2.

- Compute $\theta = \hat{\theta} + \mathbf{n}$.

Generating samples of standard deviation

The conditional density for the standard deviation is given by

$$p\,(\sigma \mid \theta,\, \mathbf{z},\, \mathbf{y}) \propto \sigma^{-N} \exp\left(-\frac{\mathbf{e}^{\mathbf{T}}\,\mathbf{e}}{2\,\sigma^2}\right) \tag{6.60}$$

where N is the sample size and p is the order of the AR model.

This is in the form of an inverse chi density [70] with N degrees of freedom. It may be shown (by differentiation) that the mode is

$$\hat{\sigma} = \left(\frac{\mathbf{e}^{\mathbf{T}} \mathbf{e}}{N} \right)^{\frac{1}{2}} \tag{6.61}$$

It is difficult to sample directly the density in equation 6.60 as it stands. The approach we will use is to generate random variates from a gamma distribution, and carry out a functional transformation to give the required distribution.

The gamma density [70] is defined as

$$p_g(x) = \frac{1}{\Gamma(\alpha)} x^{\alpha-1} \exp(-x) \tag{6.62}$$

A computer routine written in C for sampling gamma densities is provided in Numerical Recipes [95] (but it needs to be modified slightly to cope with non-integer values of α). Comparing equation 6.60 with equation 6.62 we can see that we need to set $x = \mathbf{e}^{\mathbf{T}} \mathbf{e} / 2\,\sigma^2$. Let us write

$$p_g(x)\, dx \propto p(\sigma \mid \theta, \mathbf{z}, \mathbf{y})\, d\sigma \tag{6.63}$$

It is easy to show that letting $\alpha = \frac{N-1}{2}$ will accomplish the required transformation.

The following steps will generate σ, with the required inverse chi density.

1. Generate a gamma variate with $\alpha = \frac{N-1}{2}$ degrees of freedom.

$$x_i \leftarrow x^{\alpha-1} \exp(-x)$$

2. Take the reciprocal of the square root and scale the result.

$$\sigma_i = \left(\frac{\mathbf{e}^{\mathbf{T}} \mathbf{e}}{2} \right)^{\frac{1}{2}} \frac{1}{\sqrt{x_i}}$$

6.6 Relationship between the three restoration methods

It is obvious at this stage that there are similarities between the three methods of audio restoration developed above.

6.6.1 ML vs Gibbs

Each iteration of the ML algorithm involves estimating the missing data and then estimating the autoregressive parameters. The Gibbs sampler uses exactly the same estimates for the missing data and autoregressive parameters but also adds noise to these estimates. In fact, the ML approach above is actually a special case of the Gibbs sampler with the standard deviation equal to zero. If $\sigma = 0$ then the Gibbs sampler would first compute an ML estimate of the AR parameters and add no random Gaussian perturbation, and second would compute an ML estimate of the missing data and add no random Gaussian perturbation to this either. The random drawing of the standard deviation would be completely skipped because it is equal to zero. In other words, the Gibbs sampler may be visualized as a noisy version of the ML algorithm.

6.6.2 Gibbs vs EM

As noted in section 6.4, the Gibbs sampler produces a sample from the predictive density $p(\mathbf{z} \mid \mathbf{y})$. It is interesting to note that this density is maximized by the EM algorithm. In other words, the EM interpolant is the most probable (but is not a typical) Gibbs interpolant.

6.6.3 EM vs ML

The difference between the system of equations defining the EM solution (equation 6.35) and the ML solution (equation 6.14) depends on the variance of the excitation process. The relative size of the excitation depends on the position of the poles of the AR model on the unit circle. For example, for sine waves the excitation is effectively zero because the poles of the AR model lie exactly on the unit circle. In this particular case the EM and ML interpolants will be identical. In many cases the dominant poles of the AR process are close to the unit circle, which implies that ML and EM interpolants will be very similar.

The difference between the two techniques is really due to the fact that the EM algorithm maximizes the predictive density $p(\mathbf{z} \mid \mathbf{y})$ and is thus a marginal estimator. The ML algorithm on the other hand maximizes likelihood $p(\mathbf{z}, \mathbf{y} \mid \theta)$ and thus produces a joint estimate of the AR parameters θ as well as the required missing data $\mathbf{z}$.

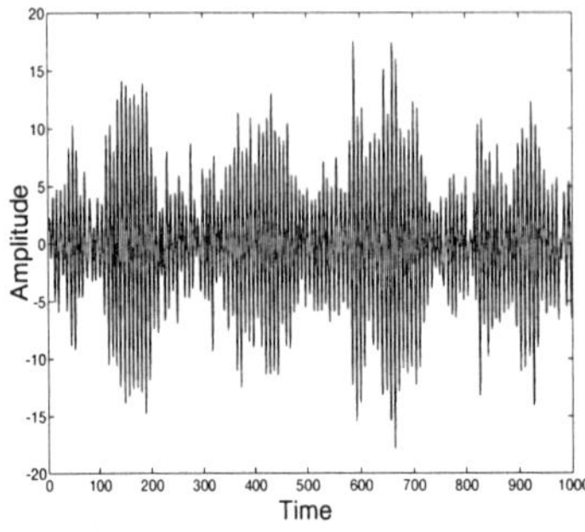

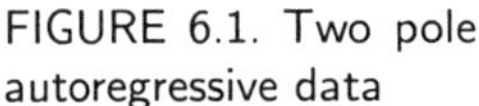
FIGURE 6.1. Two pole autoregressive data

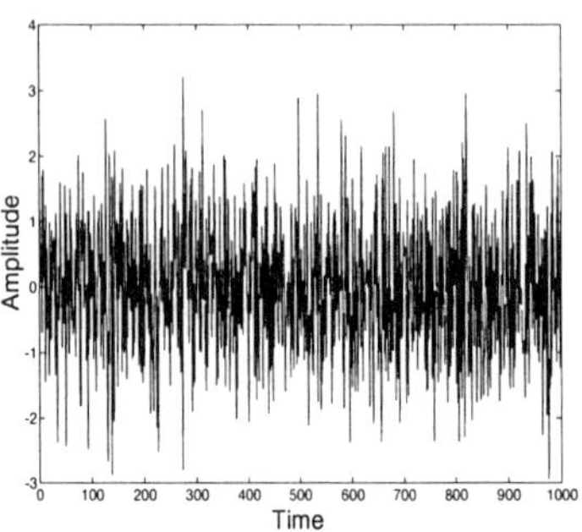

FIGURE 6.2. Excitation sequence for original data

6.7 Simulations

6.7.1 Autoregressive model with poles near unit circle

Figure 6.1 shows one thousand samples of a synthetic Gaussian AR process with poles at $0.7 + j\,0.7$ and $0.7 - j\,0.7$. The corresponding excitation sequence is shown in figure 6.2. Samples 300 to 499 were removed and subsequently replaced by Gibbs, ML and EM interpolants.

Figure 6.3 shows a sample Gibbs restoration of the missing samples after twelve iterations and figure 6.4 shows the estimated excitation sequence (formed by convolving the AR coefficients with the augmented data). It is interesting to note that the original data and the Gibbs sample restoration are not the same. They do correspond closely at the edges of the gap but are completely different in the middle.

Figure 6.5 and figure 6.7 show the ML and EM restorations respectively. The two restorations which both needed six iterations each are virtually identical. This example illustrates the classic problem of using maximum likelihood or maximum predictive density to interpolate over long gaps in an AR process. In both cases the interpolant is severely attenuated inside the gap. The overall impression obtained is that the interpolants simply do not fit.

Figure 6.6 and figure 6.8 show the estimated excitation corresponding to the ML and EM restorations respectively. It is immediately noticeable that there is little or no excitation in the gap at all. This should not surprise us too greatly. In section 6.2.1 it was shown that maximizing likelihood is equivalent to minimizing excitation energy. The missing data is therefore chosen to minimize the excitation energy. This accounts for the rather strange results.

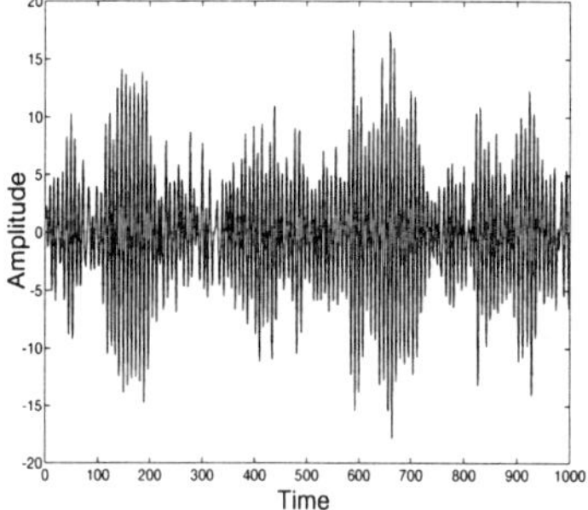

FIGURE 6.3. Gibbs restored two pole autoregressive data

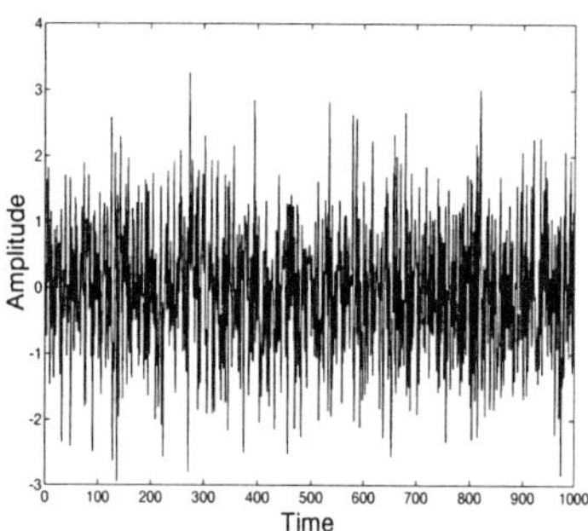

FIGURE 6.4. Estimated excitation sequence for Gibbs restoration

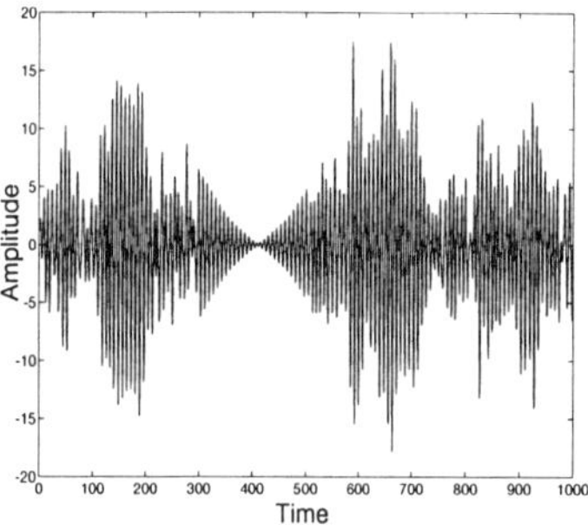

FIGURE 6.5. ML restored two pole autoregressive data

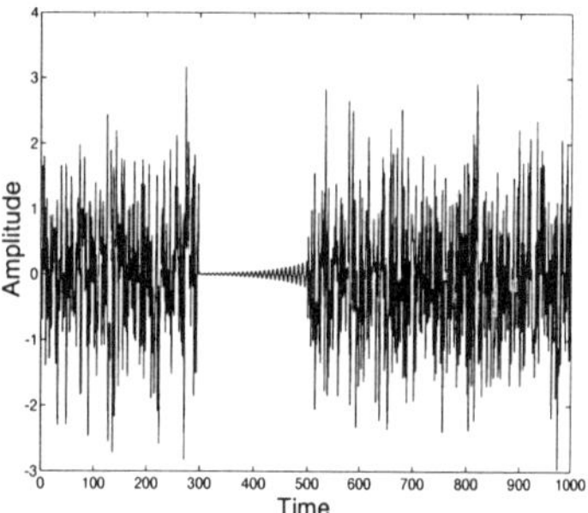

FIGURE 6.6. Estimated excitation sequence for ML restoration

6.7.2 Autoregressive model with poles near origin

Figure 6.9 shows one thousand samples of a synthetic Gaussian AR process with poles at $0.5 + j0.5$, $0.5 - j0.5$ and 0.5. The corresponding excitation sequence is shown in figure 6.10. Samples 500 to 649 were removed and subsequently replaced by Gibbs, ML and EM interpolants.

Figure 6.11 shows a sample Gibbs restoration of the missing samples and figure 6.12 shows the estimated excitation sequence. As in the previous case, the Gibbs interpolant differs from the original data in the gap.

Figure 6.13 shows the ML restoration. The EM restoration plotted in figure 6.15 is virtually identical. The attenuation of the interpolant is very severe, with the interpolant exponentially decreasing to virtually zero after just a few samples.

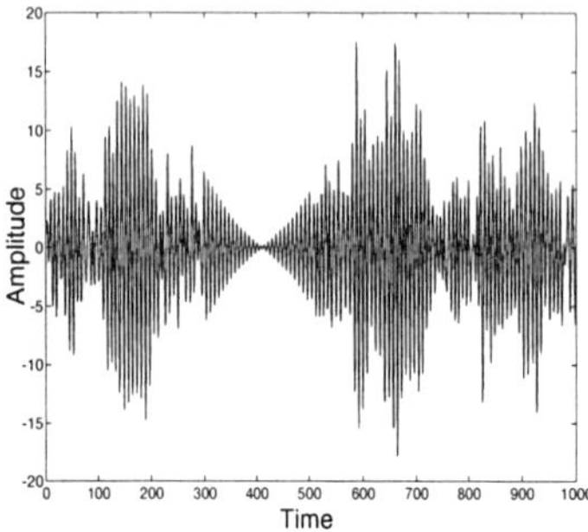

FIGURE 6.7. EM restored two pole autoregressive data

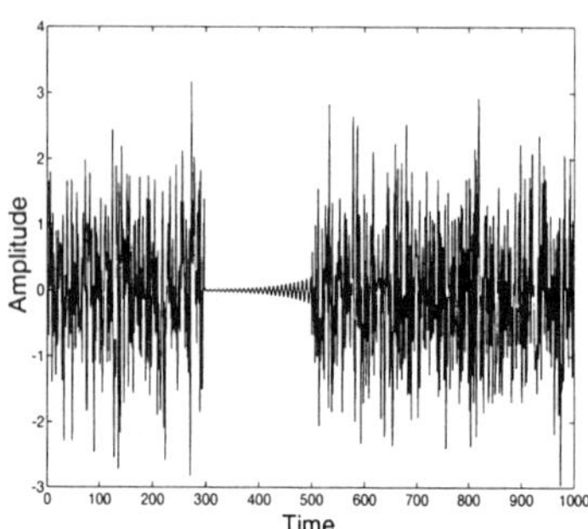

FIGURE 6.8. Estimated excitation sequence for EM restoration

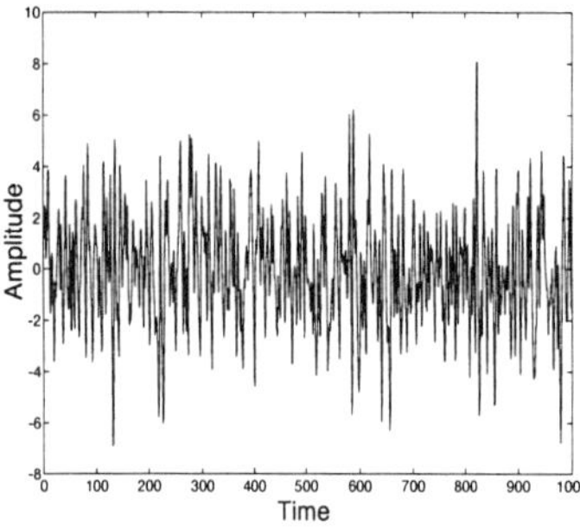

FIGURE 6.9. Three pole autoregressive data

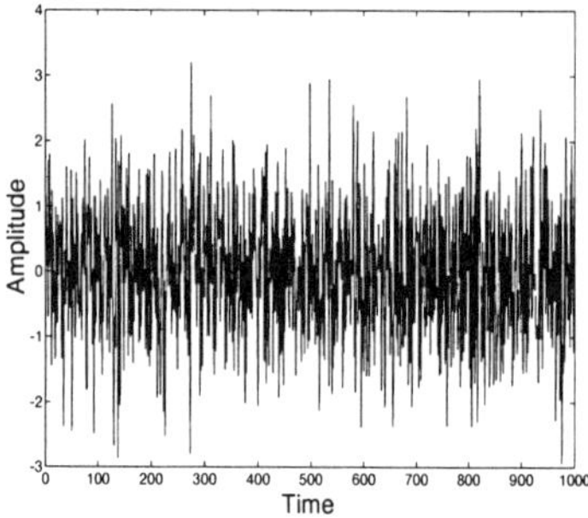

FIGURE 6.10. Excitation sequence for three pole autoregressive data

Figure 6.14 and figure 6.16 show the estimated excitation corresponding to the ML and EM restorations respectively, and as in the previous example, that there is little or no excitation energy in the gap.

The marked difference between the results for the first example data set plotted in figure 6.1 and those plotted in figure 6.9 is due to the positions of the poles of the AR process. In figure 6.1 the poles were deliberately chosen to be close to the unit circle. This means that the transients set up by the excitation sequence are slowly decaying sinusoids. These transients are sufficiently long lived that the interpolant can almost span the gap. In figure 6.9 the poles were chosen to be far away from the unit circle. Therefore the transients set up by the noise excitation are short lived decaying

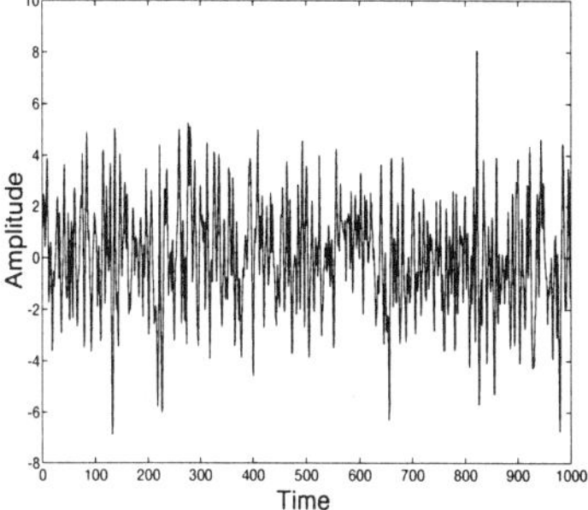

FIGURE 6.11. Gibbs restored three pole autoregressive data

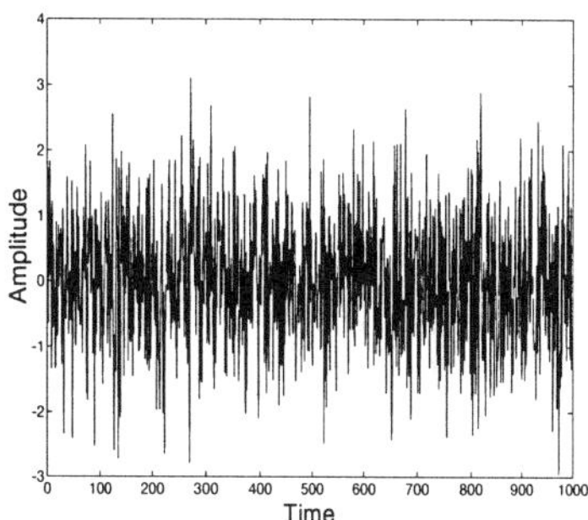

FIGURE 6.12. Estimated excitation sequence for Gibbs restoration

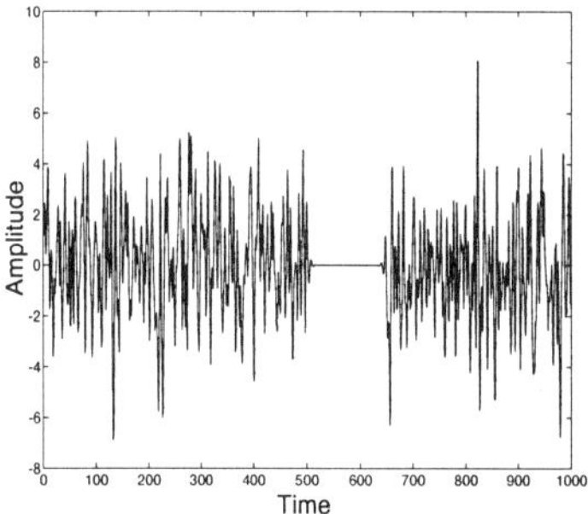

FIGURE 6.13. ML restored three pole autoregressive data

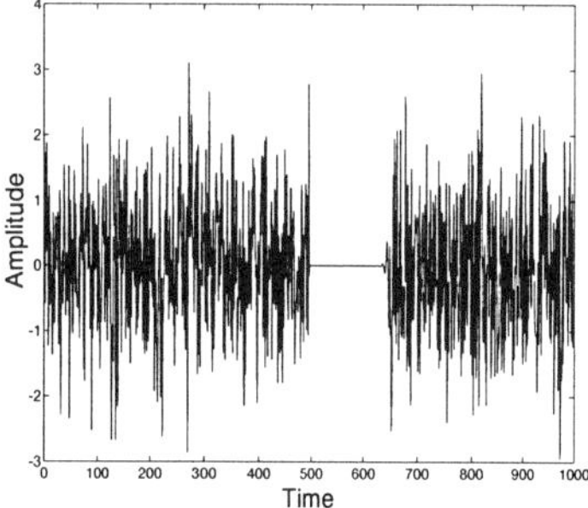

FIGURE 6.14. Estimated excitation sequence for ML restoration

sinusoids, which very rapidly decay to give an interpolant that is effectively zero valued in the gap.

This example is also interesting because it gives the biggest difference between the EM and ML interpolants. As noted earlier in section 6.3 the difference between the ML and EM interpolants depends on the magnitude of the excitation relative to the observed data. This ratio depends on the position of the poles of the AR process, being zero if the poles lie on the unit circle and unity if the poles lie on the origin (i.e. a pure delay). Since in this case the poles lie relatively close to the origin then the difference between ML and EM interpolants will be relatively large. The difference between the ML and EM restorations in figure 6.13 and figure 6.15 respectively is plotted in figure 6.17.

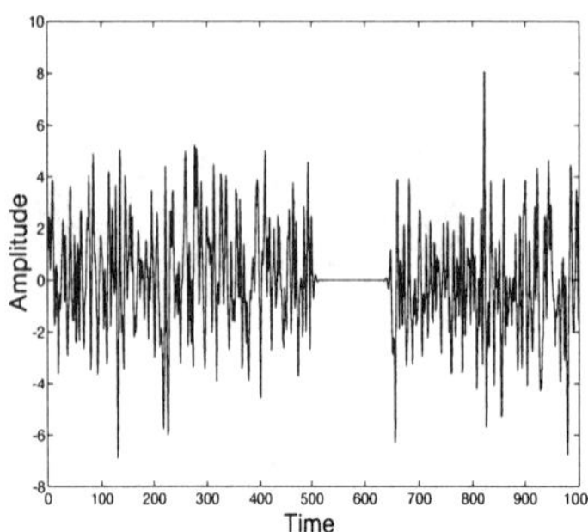

FIGURE 6.15. EM restored three pole autoregressive data

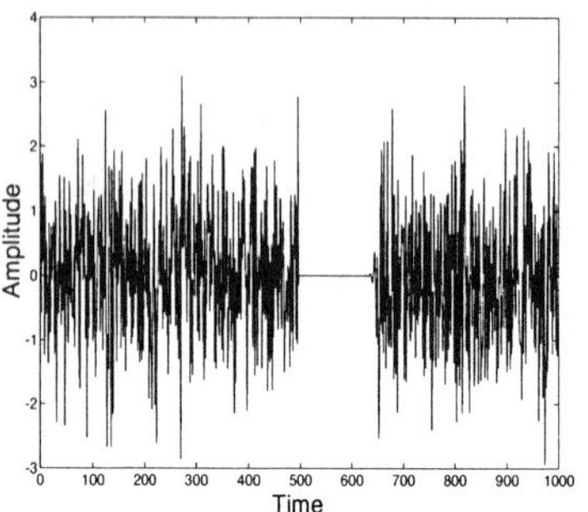

FIGURE 6.16. Estimated excitation sequence for EM restoration

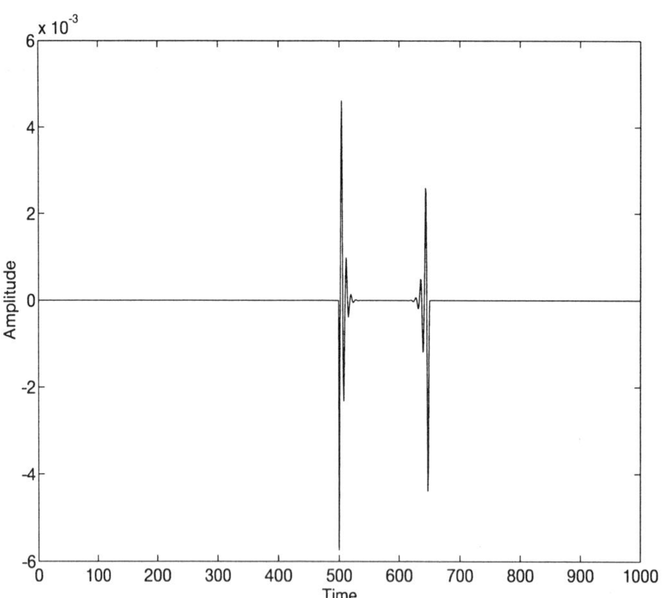

FIGURE 6.17. Difference between EM and ML interpolants

The interesting point that may be gleaned from this discussion is that ML and EM give different results because they are each answering different questions. However, in most cases the difference between the results obtained using each method is insignificant.

6.7.3 Sine wave

Consider the data set plotted in figure 6.18 which shows one thousand samples of a sine wave of unit amplitude and with period equal to two hundred sample intervals. This is a special case of an AR process. There are two poles lying on the unit circle and the AR process is marginally stable. For a pure sine wave of the kind plotted in figure 6.18 the excitation is exactly zero. The reason why the data is non-zero is because at some time before the data were observed a unit impulse excited the modes of the AR filter. Because of the position of the poles, the resulting transients have zero exponential decay and therefore continue indefinitely in time.

This data set provides a very important test for the Gibbs sampler and for the ML and EM algorithms. Although the signal may be described as an AR process it is also completely deterministic. First, it is expected that all three methods devised for interpolating the missing data should converge to the unique "correct" restoration that is known to exist. Second, there is no excitation at all so all restorations (including Gibbs) should have no excitation in the gap.

A sample interpolant obtained after fifty iterations of a Gibbs sampler is shown in figure 6.20. The excitation sequence plotted in figure 6.21 is non-zero — but the amplitude ($\approx 10^{-6}$) is negligible. As will be seen later, it can be further reduced by carrying out more iterations of the Gibbs sampler.

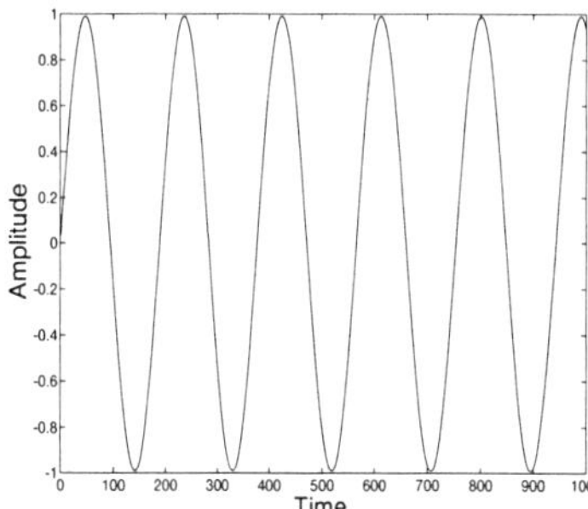

FIGURE 6.18. Sine wave data

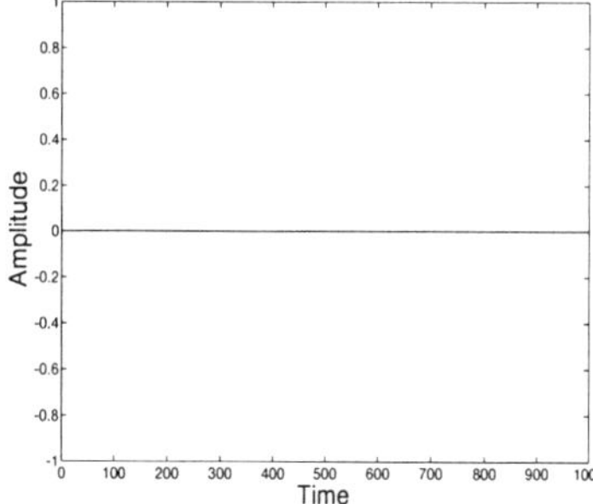

FIGURE 6.19. Excitation sequence for sine wave data

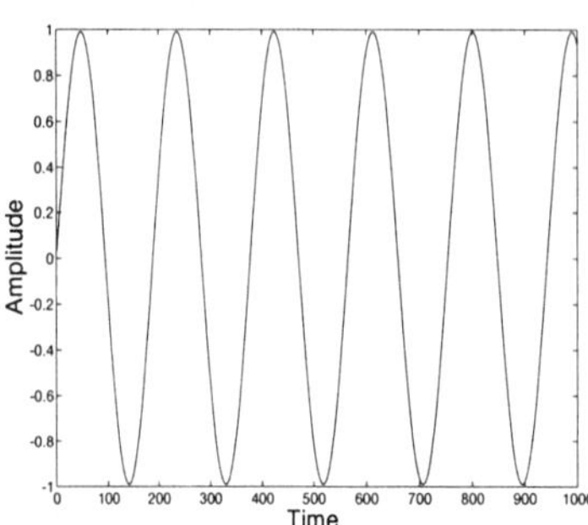

FIGURE 6.20. Gibbs restored sine wave data

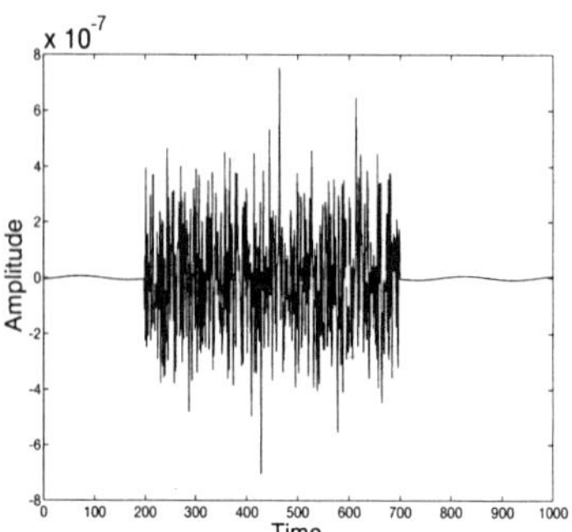

FIGURE 6.21. Estimated excitation sequence for Gibbs restoration

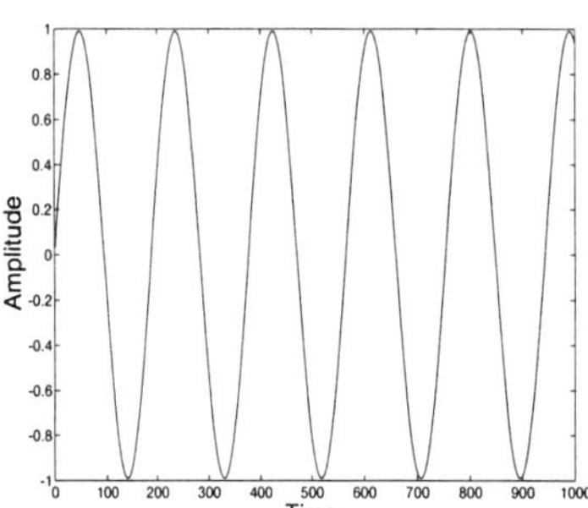

FIGURE 6.22. ML restored sine wave autoregressive data

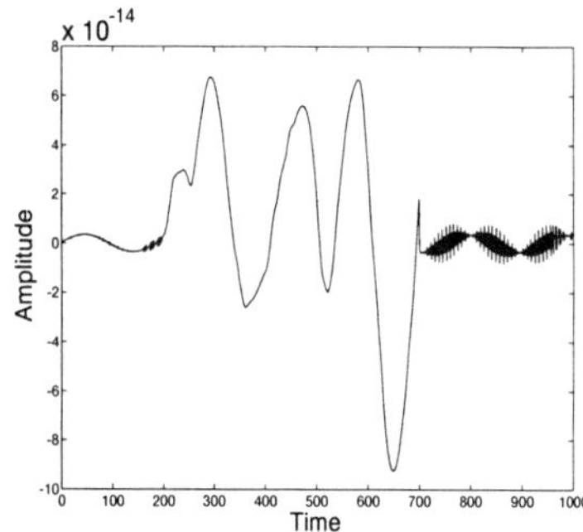

FIGURE 6.23. Estimated excitation sequence for ML restoration

The ML interpolant obtained after fourteen iterations is plotted in figure 6.22. Again, as for the Gibbs sampler, the signal reconstruction is near perfect. The estimated excitation plotted in figure 6.23 attains a maximum amplitude of less than 10^{-13}.

The EM interpolant is shown in figure 6.24 and its corresponding excitation is shown in figure 6.25. The excitation in this case is of the order of 10^{-11}. This is the greatest accuracy that could be attained given the finite precision of the computer used to generate the results.

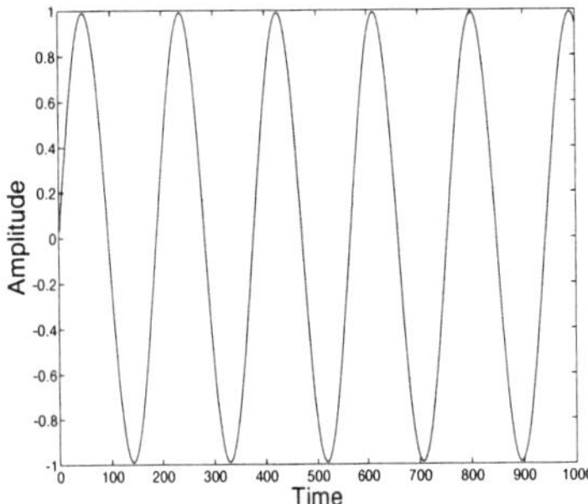

FIGURE 6.24. EM restored sine wave autoregressive data

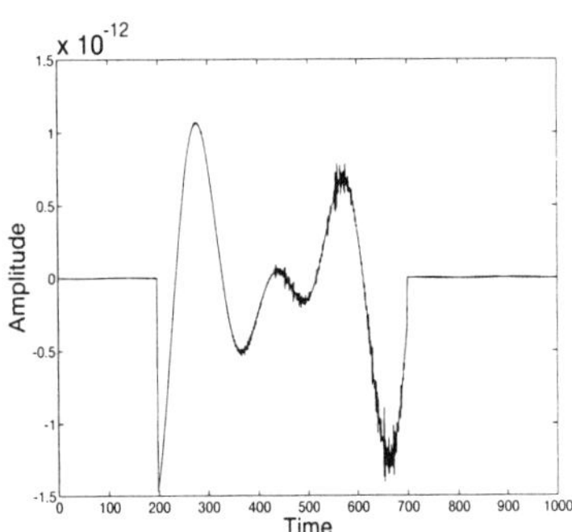

FIGURE 6.25. Estimated excitation sequence for EM restoration

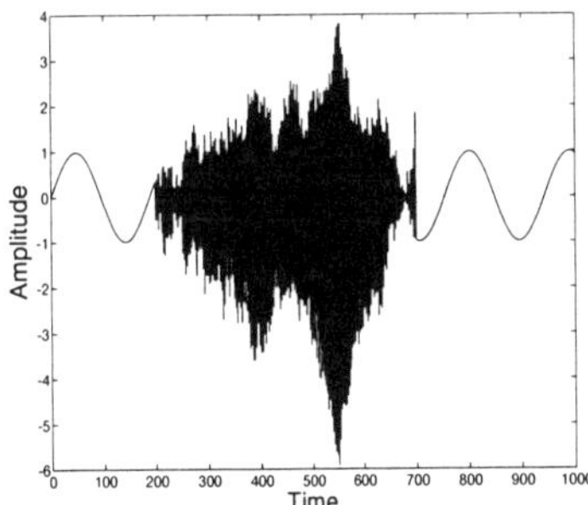

FIGURE 6.26. Gibbs sample restoration: one iteration

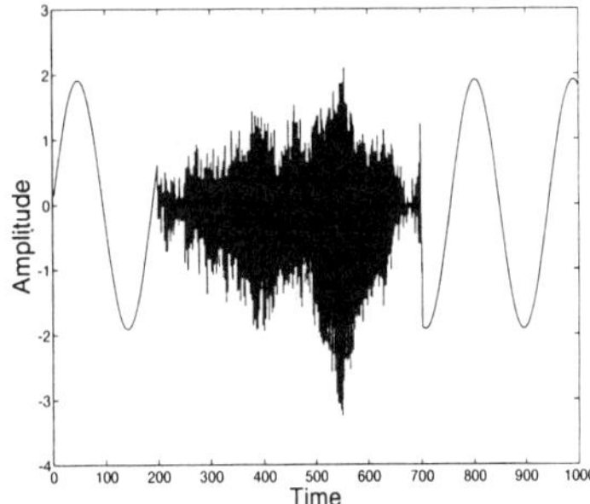

FIGURE 6.27. Gibbs sample excitation: one iteration

6.7.4 Evolution of sample interpolants

In this section, we examine the evolution of the Gibbs sample interpolants over a number of iterations. The example chosen is the sine wave of the previous section. Let us begin with a zero interpolant.

After one iteration, the sample interpolant as shown in figure 6.26 is very poor. The estimated excitation plotted in figure 6.27 does not even vaguely resemble a white i.i.d. Gaussian noise process.

After ten iterations, the interpolant plotted in figure 6.28 contains less high frequency components and the corresponding excitation plotted in figure 6.29 has decreased by an order of magnitude.

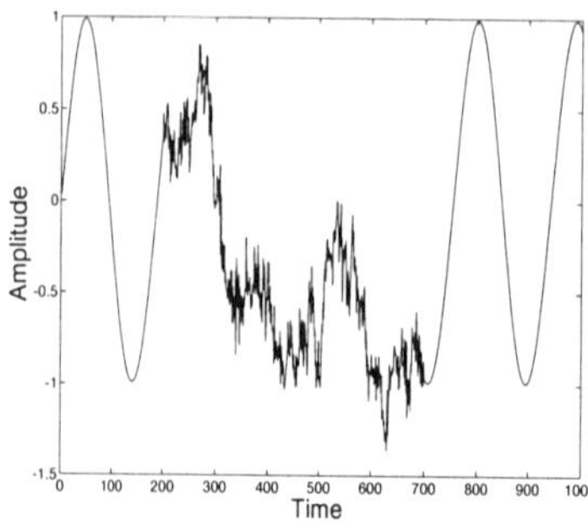

FIGURE 6.28. Gibbs sample restoration: ten iterations

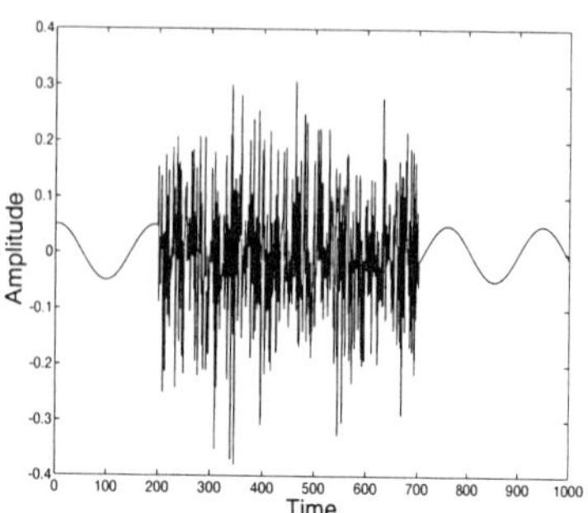

FIGURE 6.29. Gibbs sample excitation: ten iterations

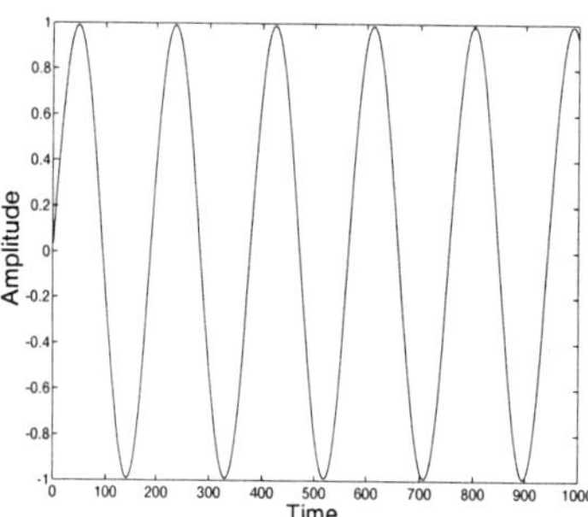

FIGURE 6.30. Gibbs sample restoration: fifty iterations

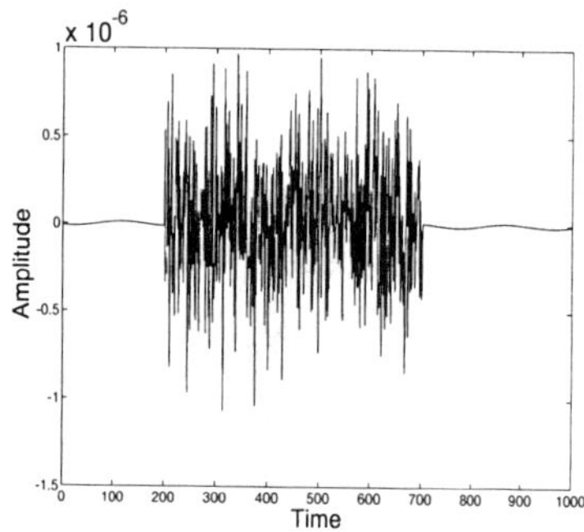

FIGURE 6.31. Gibbs sample excitation: fifty iterations

The process is continued until iteration 50 as shown in figure 6.30. The excitation amplitude decreases by a factor of twenty approximately after every ten iterations.

After thirty iterations a plausible interpolant has appeared, whilst after forty iterations the interpolant has, for all intents and purposes, converged.

The most remarkable feature of this example is the steady monotonic decrease of the magnitude of the excitation from about 4 after the first iteration to a magnitude of less than 10^{-6} after fifty iterations.

The rate of convergence in this example is admittedly painfully slow. The chief factor determining the rate of convergence is the number of missing data. In this case, 50% of the data are missing which leads to an abnormally slow rate of convergence. The same can be said of the ML and EM algorithms. A theoretical treatment of the subject is presented by Tanner [126] and Dempster, Laird and Rubin [29] in the context of the convergence of the EM algorithm in general. Veldhuis [137] also discusses the convergence of the ML algorithm (which he presents as a Newton-Raphson optimization) for the interpolation of autoregressive processes.

6.7.5 Hairy sine wave

In this section, we restore the data plotted in figure 6.32 which shows one thousand samples of a sine wave of unit amplitude with added white Gaussian noise of standard deviation 0.2.

Five hundred samples were removed starting with sample number 250 and were restored using the Gibbs, EM and ML algorithms. An AR model of order 80 is assumed. The rather large model order is due to the fact that the data is not exactly describable by a finite length AR model. In fact the data is a realization of an ARMA process [22, 114]. All ARMA models can be expressed as AR models of infinite order, but can be well approximated [22] by AR models if the order of the model is sufficiently high. This is fortunate since ARMA models (unlike AR models) are difficult [22] to work with.

The EM and ML restorations plotted in figure 6.36 and figure 6.38 are extremely interesting. The interpolant is smooth which indicates that some form of noise filtering is taking place. In this case there is no appreciable attenuation of the interpolant towards the centre of the gap.

Note that the excitation plotted in figures 6.33, 6.35, 6.37 and 6.39 begins at sample number 81. This is because of the nature of the AR model which in this case means that each sample excitation may be expressed in terms of the previous 80 sample data. We have no knowledge of the data before the first sample; therefore only the estimated residuals from sample number 81 can be given.

The eighty poles of the AR model are distributed over the unit circle, with some lying closer to the unit circle than others. The interpolant spans the gap with almost no attenuation which means that at least one pair of poles are positioned extremely close to the unit circle, with the result that these poles dominate the interpolant towards the centre of the gap.

The Gibbs sample interpolant, on the other hand, looks perfectly plausible. Compared to EM and ML, the Gibbs sampler produces a "hairy" interpolant with a texture that is consistent with the data on either side of the gap. The EM and ML interpolants are "bald" and are obviously inconsistent with the data.

6.7.6 Real data: Tuba

Figure 6.40 shows one thousand samples of a piece of tuba music. The sampling rate is 44.1 kHz so the section of music is just less than 22.7 ms in duration. Because of the very short duration of the signal it is usually a

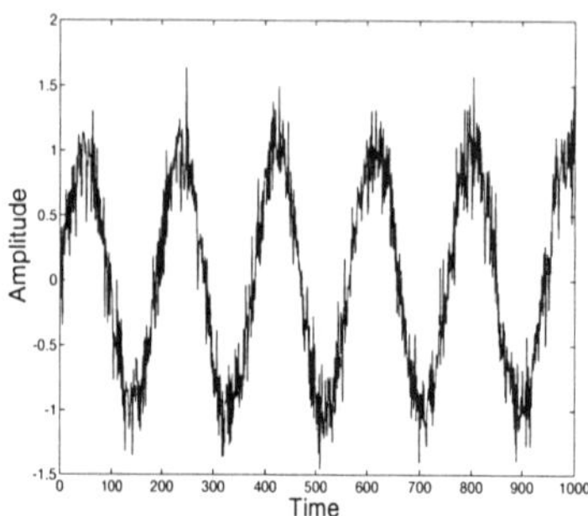

FIGURE 6.32. Hairy sine wave data

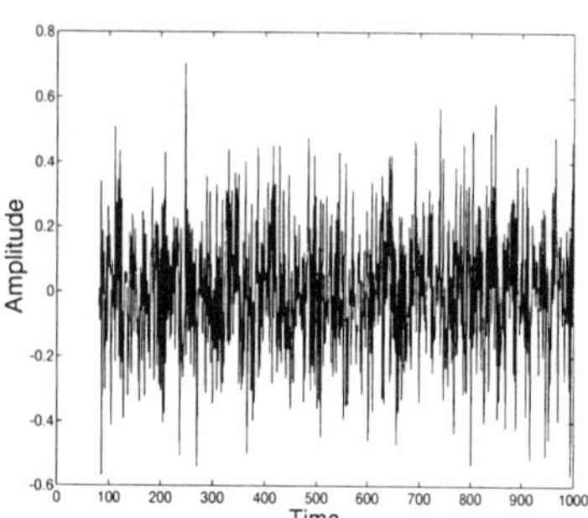

FIGURE 6.33. Estimated excitation sequence for hairy sine wave data

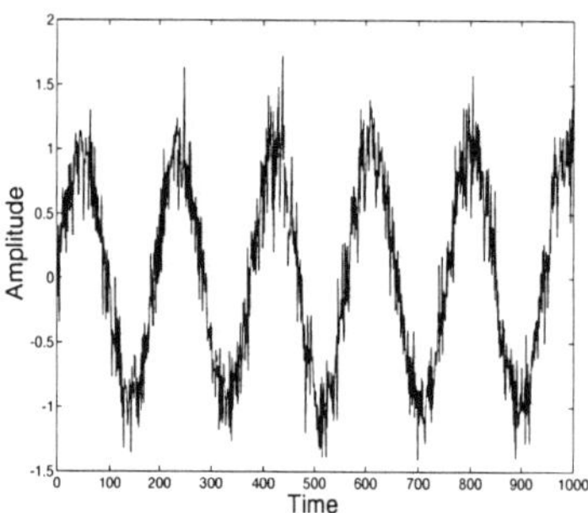

FIGURE 6.34. Gibbs restored hairy sine wave data

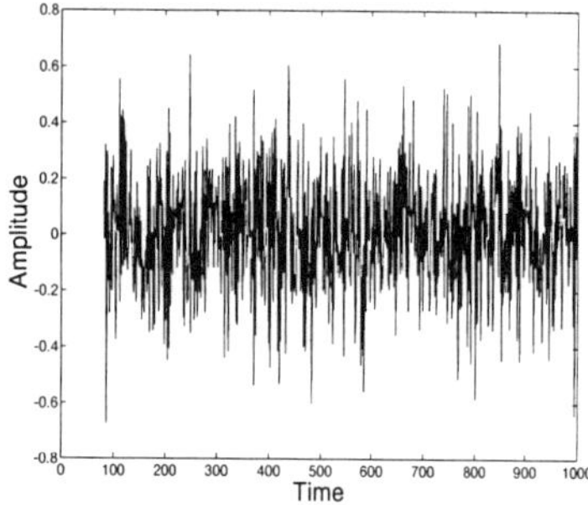

FIGURE 6.35. Estimated excitation sequence for Gibbs restoration

reasonable assumption [136] to treat the data as a stationary autoregressive process. Also in common with other work on audio signals [40, 136] we shall assume that the order is 40.

The estimated excitation sequence for the original data is plotted in figure 6.41. The excitation sequence appears to be white i.i.d. stationary Gaussian noise which would support the choice of signal model. To provide a severe test of the interpolation techniques two hundred samples were removed to be later restored by interpolation.

A Gibbs sample interpolant is plotted in figure 6.42 and the corresponding excitation sequence is plotted in figure 6.43. The excitation sequence appears to be i.i.d. Gaussian noise.

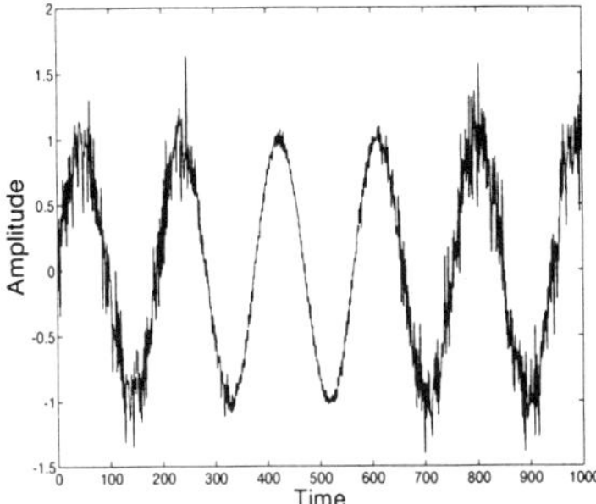

FIGURE 6.36. ML restored hairy sine wave data

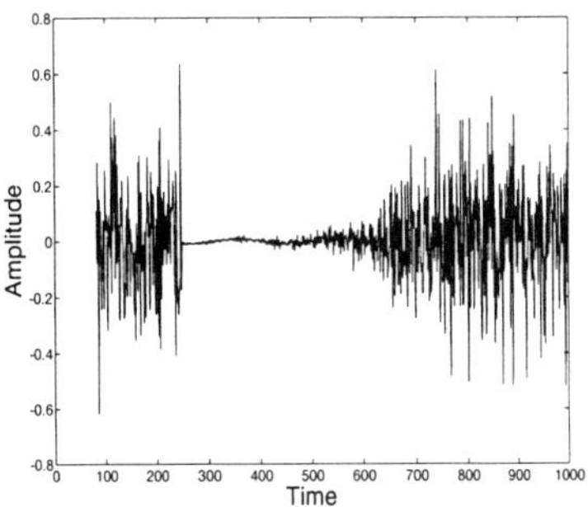

FIGURE 6.37. Estimated excitation sequence for ML restoration

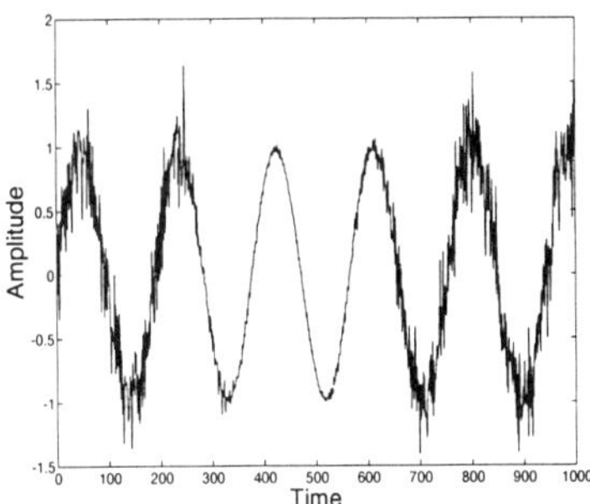

FIGURE 6.38. EM restored hairy sine wave data

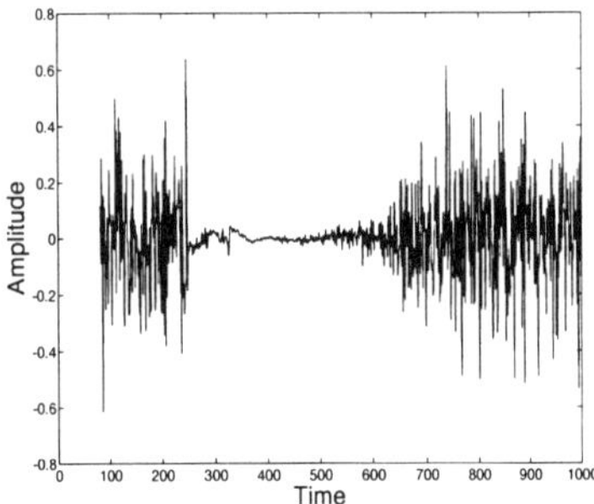

FIGURE 6.39. Estimated excitation sequence for EM restoration

The ML interpolant plotted in figure 6.44 has a noticeable "dead zone" where the signal amplitude level drops to almost zero at the centre of the gap. The corresponding excitation sequence plotted in figure 6.45 explains the sudden drop.

The EM interpolant plotted in figure 6.46 is also disappointing. The excitation plotted in figure 6.47 is slightly different to that of the ML restoration, but the "dead zone" is just as pronounced as before.

In order to provide a subjective assessment of the restoration algorithms listening tests were carried out and the performance of the three techniques was compared. The ML and EM algorithms (not surprisingly) sound identical. There is noticeable distortion with the gap in the data sounding like a

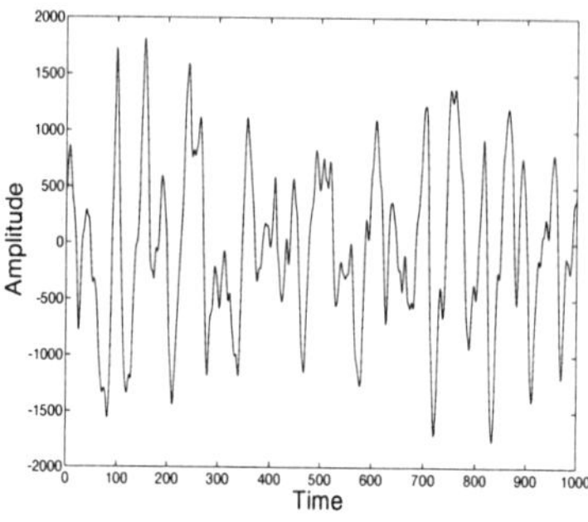

FIGURE 6.40. Tuba data

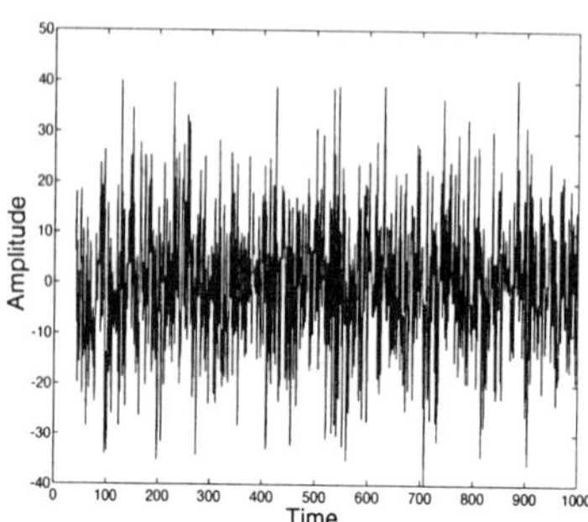

FIGURE 6.41. Excitation sequence for tuba data

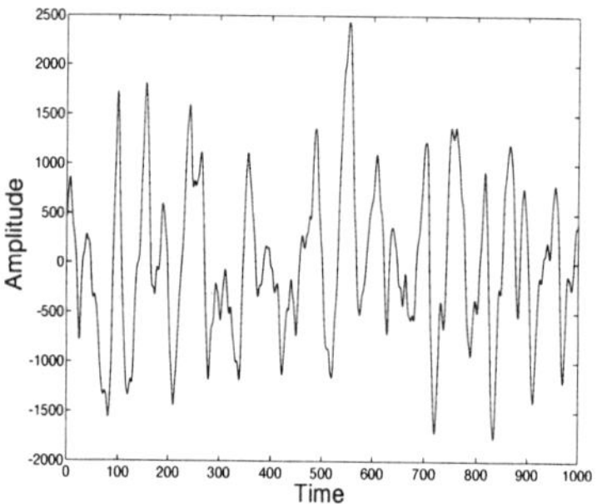

FIGURE 6.42. Gibbs restored tuba data

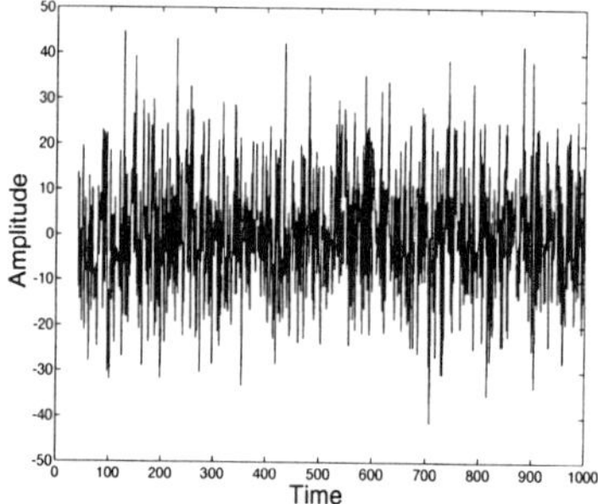

FIGURE 6.43. Estimated excitation sequence for Gibbs restoration

slight thump. No distortion can be discerned in the Gibbs sampler restoration.

6.7.7 Real data: Sinéad O'Connor

Figure 6.48 shows one thousand samples of a piece of music recorded from "Nothing Compares 2U" by Sinéad O'Connor. The sampling rate is 44.1 kHz as in the case of the tuba above. The data is modelled as a stationary autoregressive process of order 40. The estimated excitation sequence for the original data is plotted in figure 6.49.

This example is quite different to that of the tuba data for two reasons. First, it is a piece of modern pop music that has quite a different sound to

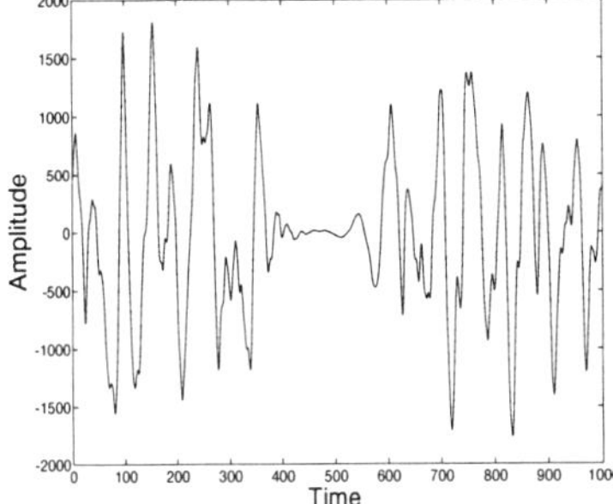

FIGURE 6.44. ML restored tuba data

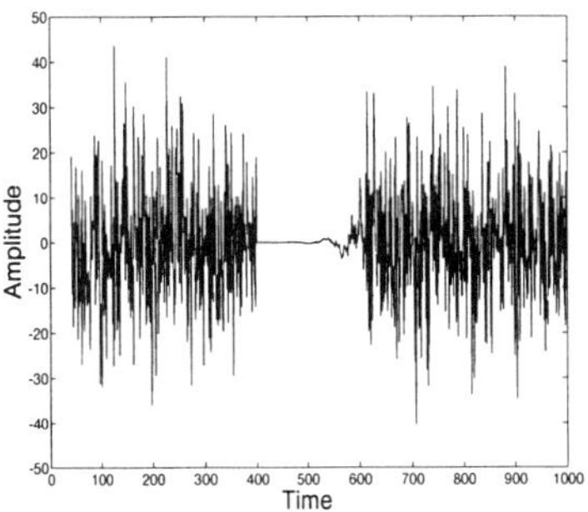

FIGURE 6.45. Estimated excitation sequence for ML restoration

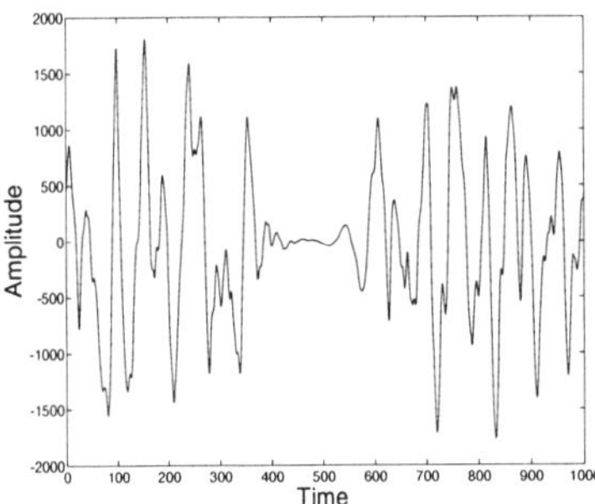

FIGURE 6.46. EM restored tuba data

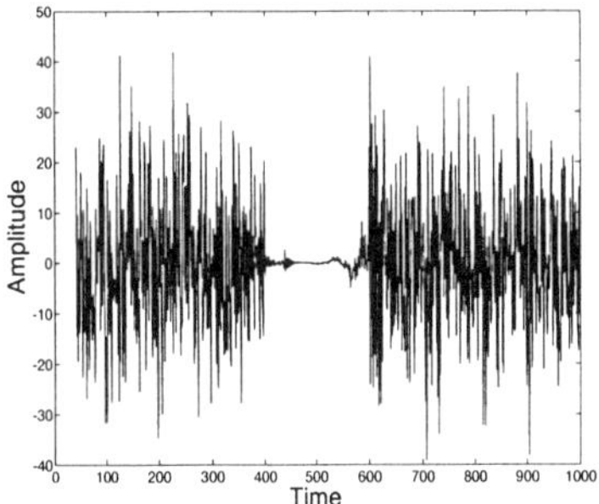

FIGURE 6.47. Estimated excitation sequence for EM restoration

the classical music above. More importantly, it is a sample of human singing voice which should be considerably different in nature to the resonant tones produced by a musical instrument.

To provide a severe test of the interpolation techniques two hundred samples were removed to be later restored by interpolation.

A Gibbs sample interpolant is plotted in figure 6.50 and the corresponding excitation sequence is plotted in figure 6.51. The excitation sequence appears to be i.i.d. Gaussian noise.

The ML interpolant plotted in figure 6.52 has a noticeable "dead zone" where the signal amplitude level is attenuated at the centre of the gap but there is still some residual oscillation. The corresponding excitation sequence is plotted in figure 6.53.

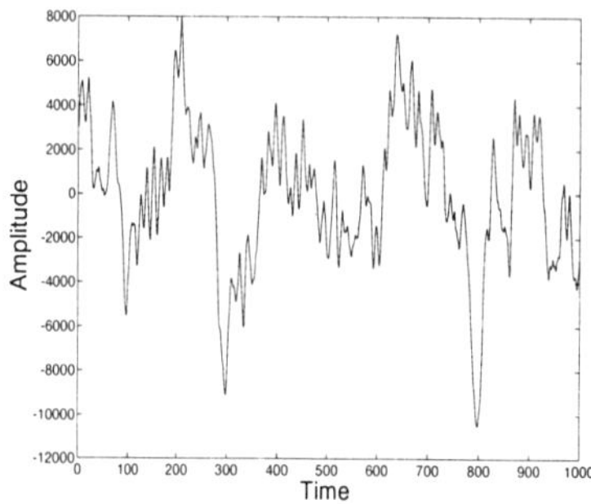

FIGURE 6.48. Sinéad O'Connor in action

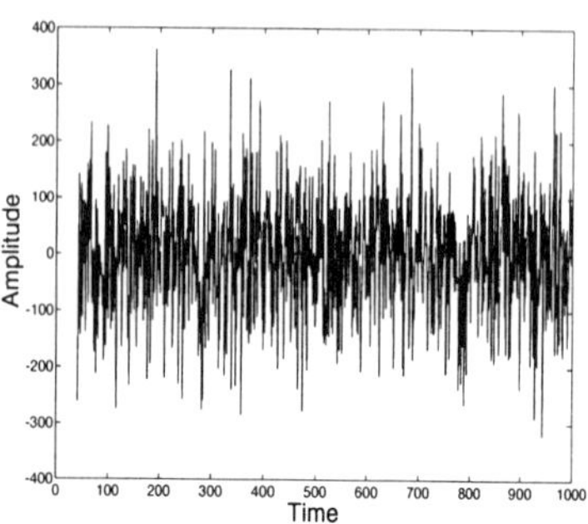

FIGURE 6.49. Excitation sequence for Sinéad O'Connor

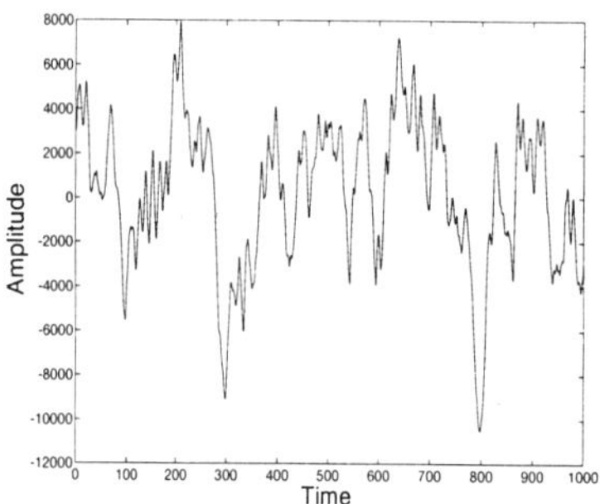

FIGURE 6.50. Gibbs restoration Sinéad O'Connor

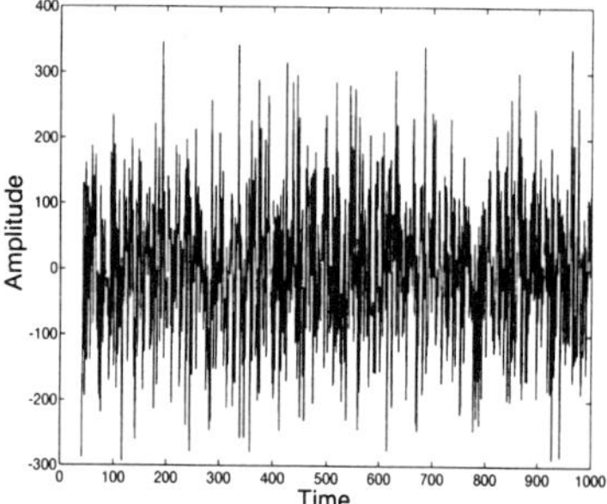

FIGURE 6.51. Estimated excitation sequence for Gibbs restoration

The EM interpolant plotted in figure 6.54 is practically the same as the ML interpolant. This is despite there being some slight difference in the excitation plotted in figure 6.55 compared with that of figure 6.53.

Listening tests were carried out and the performances of the three techniques were compared. As before, the ML and EM algorithms sound identical. The gap in the data produces a distortion which sounds like a glottal stop or a "ng" swallowing sound that is quite noticeable in a singing voice. As before, no distortion at all can be discerned in the Gibbs sampler restoration.

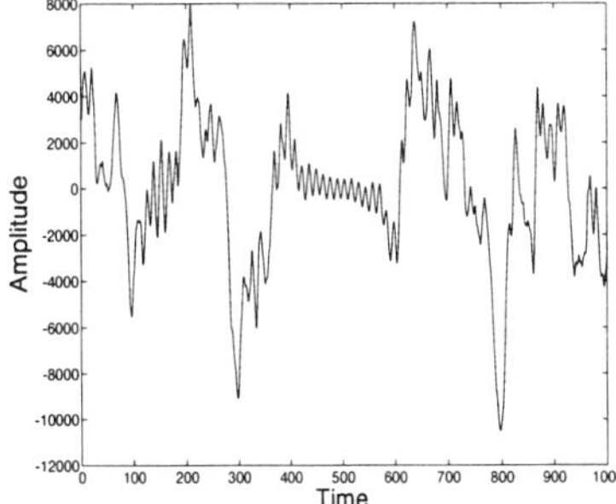

FIGURE 6.52. ML restoration of Sinéad O'Connor

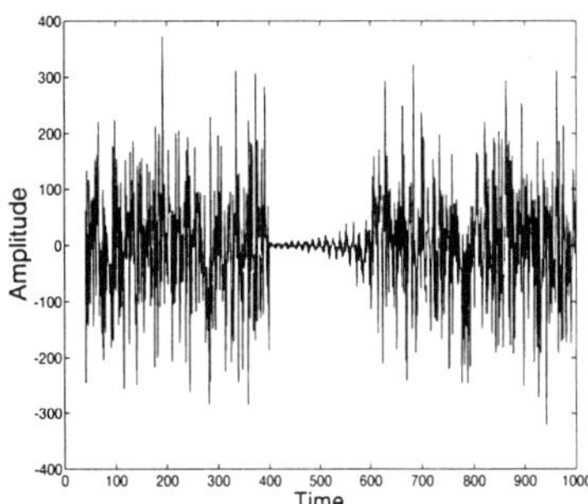

FIGURE 6.53. Estimated excitation sequence for ML restoration

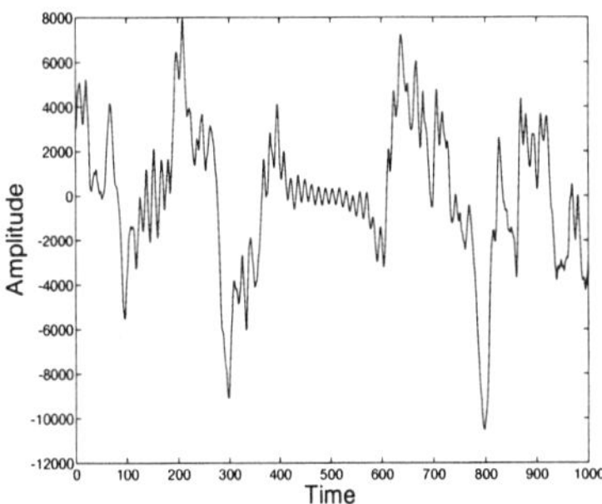

FIGURE 6.54. EM restoration of Sinéad O'Connor

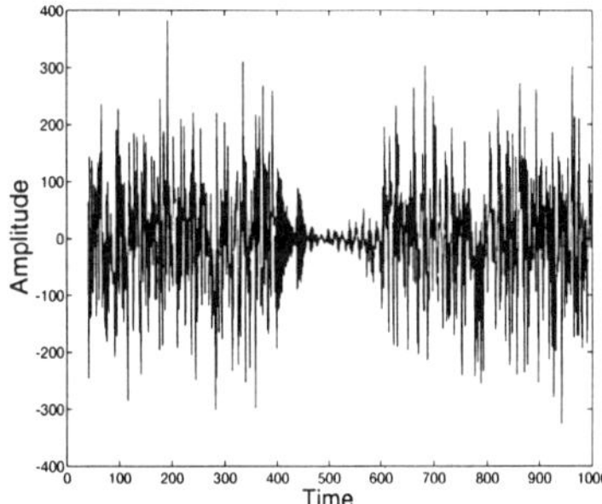

FIGURE 6.55. Estimated excitation sequence for EM restoration

6.8 Discussion

6.8.1 *The temperature of an interpolant*

In this section, ideas from the field of thermodynamics are used to justify why the Gibbs sample interpolant is a typical interpolant.

Consider the Boltzmann (or canonical) distribution which is given by:

$$p_B(E) = \exp\left[-\frac{E}{T}\right] \tag{6.64}$$

where T is the temperature and E is the energy of a given state of the thermodynamical system. For convenience, the Boltzmann constant is taken to be unity.

In nature, there is a tendency to minimize energy whereas for estimation one would like to maximize probability. Let us make the substitution

$$E(\omega) = -\log\left[p(\omega)\right] \tag{6.65}$$

where ω denotes the thermodynamic state of the system, or equivalently, the parameter vector.

Therefore the Boltzmann distribution may be written as:

$$p_B\left[E\right] = \exp\left[\frac{\log\left(p(\omega)\right)}{T}\right] \tag{6.66}$$

At low temperature, samples from the Boltzmann distribution will tend to be of low energy or, equivalently, high probability density. At the ambient temperature, $T = 1$, samples from the Boltzmann distribution will be drawn from the probability density $p(\omega)$.

In fact, we now have a neat way to compare the different restoration techniques.

- The EM algorithm samples the Boltzmann distribution at $T = 0$ with $E = -\log\left[p(\mathbf{z} \mid \mathbf{y})\right]$
- The ML solution samples the Boltzmann distribution at $T = 0$ with $E = -\log\left[p(\mathbf{z}, \theta \mid \mathbf{y})\right]$
- But Gibbs samples the Boltzmann distribution at $T = 1$ with $E = -\log\left[p(\mathbf{z} \mid \mathbf{y})\right]$

In other words, the Gibbs sampler produces an interpolant that has the *same temperature* as the observed data. In fact, as noted by Neal [82] the Gibbs sampler is sometimes called the *heatbath algorithm*!

6.8.2 Data augmentation

To summarise, the Gibbs restoration is a random vector drawn from the predictive density $p(\mathbf{z} \mid \mathbf{y})$. Although the Gibbs sampler actually draws samples from $p(\mathbf{z}, \theta, \sigma \mid \mathbf{y})$, the autoregressive parameters θ and standard deviation σ are ignored which means that $\mathbf{z}$ is being simulated from the marginal (or predictive) density $p(\mathbf{z} \mid \mathbf{y})$. This point is discussed by Tanner [126].

The problem with the Gibbs sampling scheme is that it is possible that there are situations where it may not perform well. For each sample of $\mathbf{z}$ there is just one corresponding sample of θ and σ. If either θ or σ samples are unrepresentative of the joint density, in other words the samples are drawn from the tails, then it is possible that $\mathbf{z}$ may also be unrepresentative of the marginal density.

Tanner [126] and Tanner and Wong [127] introduced an approach to sampling called data augmentation to take account of this fact. For our example, for each sample of $\mathbf{z}$, Tanner would *impute* multiple samples of θ and σ, gradually increasing the number of imputations in accordance with the number of iterations. The next sample of $\mathbf{z}$ would then be drawn from a Gaussian mixture density, with each component of the Gaussian mixture depending on an imputed value of θ and σ (conditional on the previous value of $\mathbf{z}$). Tanner and Wong [127] also argue that the convergence properties of the iterative sampling scheme are improved by such multiple imputations.

It would seem therefore that data augmentation is preferable to Gibbs sampling for audio restoration. The greatest difficulty with data augmentation is that one now has to draw from a uniform Gaussian mixture density to carry out data augmentation rather than from a simple Gaussian distribution as for the Gibbs sampler. At first sight, this would appear not to be a problem since one could implement the simulation of mixture variates by choosing a Gaussian component from the mixture at random and simulating from it. A little thought [82] shows that this solution is exactly equivalent to using a Gibbs sampler.

Tanner [126] and Tanner and Wong [127] also note that it is possible to sample approximately from the marginal density by using the following approximations [131] to the predictive density:

PMDA 1 : First order "profile" approximation.

$$p(\mathbf{z} \mid \mathbf{y}) \propto p(\mathbf{z}, \hat{\theta}, \hat{\sigma} \mid \mathbf{y})$$

PMDA 2 : Second order "Laplace" approximation.

$$p(\mathbf{z} \mid \mathbf{y}) \propto p(\mathbf{z}, \hat{\theta}, \hat{\sigma} \mid \mathbf{y}) \exp\left[-\frac{1}{2}(\mathbf{z} - \hat{\mathbf{z}})^{\mathbf{T}} \mathbf{C}^{-1} (\mathbf{z} - \hat{\mathbf{z}})\right]$$

where $\hat{\theta}$ and $\hat{\sigma}$ are the maximizers of the joint posterior density subject to the value of $\mathbf{z}$ and where $\mathbf{C}$ is the covariance matrix[1] of a Gaussian approximation [131] to the posterior density. Neither of these approaches are iterative like Gibbs sampling or data augmentation. PMDA stands for "Poor Man's Data Augmentation" and is designed for use by those unable to afford the expense of a full data augmentation analysis [126]. In fact, PMDA 1 has been developed independently by Rayner and Godsill [103] as a possible solution to the loss of excitation in the gap.

---

[1]This may be obtained by inverting the Hessian matrix of the log density at the conditional mode.

6.9 Concluding remarks

6.9.1 *Typical interpolants*

The key difference between Gibbs sampling and both ML and EM is in the treatment given to unknown parameters. Both ML and EM estimate the unknown parameters from the model statistics. Rayner and Godsill [103] point out that EM and ML produce interpolants that are atypical, in the sense that they do not appear to agree with the observed data. First, the interpolant is smooth which indicates that some form of noise filtering is taking place, and second, the amplitude of the interpolant decreases towards the centre of the gap. The reason for this phenomenon is that the maximum probability criterion (as used by EM and ML) minimizes excitation energy. The net result is that the excitation suddenly falls to almost zero in the gap – as illustrated by figure 6.13. The severity of this effect is dependent on the pole positions of the AR model. If poles are close to the unit circle then transients in the audio signal associated with each sample excitation are slowly decaying sinusoids. If the decay is small enough then the (smooth) interpolant will span the gap. For poles close to the origin, on the other hand, the transients set up by the excitation sequence are short lived and will therefore not span the gap.

The Gibbs sampling approach is fundamentally different to either ML or EM because it randomly simulates the unknown quantities from the model statistics. As we have seen, the result is that the Gibbs sampler produces an interpolant that is typical of the random process that is assumed to describe the data. Using random simulation is especially reasonable in view of the fact that if the data itself is modelled as being random then why should we not also model the interpolant itself as being random? A frequentist would probably agree with the fact the interpolant should be random, but would not agree with the standard deviation and the autoregressive parameters being random as well. Treating unknowns as random variables is anathema [55] to the orthodox statistical community. A happy compromise is provided by Rayner and Godsill's PMDA 1 algorithm [103] which fixes the autoregressive parameters and standard deviation in order to generate a random interpolant.

No apologies are made for randomly simulating all three unknowns. The range of possible results for the random variates follows the statistics of the posterior density. This means that the statistics of the random variates agree with the distribution describing the state of knowledge about their value.

In order to provide a subjective assessment of the work above, listening tests were carried out on real music sampled at 44.1 kHz. Artificial clicks with long duration were added to the music and were subsequently removed. The results were excellent in that no trace of the original click could be discerned. The comparison between the performance of the Gibbs sampler and that of the ML and EM algorithms was extremely favourable.

6.9.2 Computation

In this chapter, considerable emphasis was placed on the importance of efficiency and the need for good numerical methods for implementing the audio restoration techniques. It is clear that improvements can be made on the numerical methods used above. For example, a version of Levinson's algorithm suitable for use on band diagonal systems would be very useful in the missing data stage of all three restoration methods discussed in this chapter. In addition, the band diagonal Toeplitz structure of the matrix in the missing data stage is actually a convolution in disguise. This would suggest that frequency domain techniques could be used instead of time domain based matrix methods, which should give a further increase in speed, especially when the order of the AR model being used is quite large.

All the restoration methods described have been iterative. In the case of the ML and EM algorithms the convergence was towards a unique *optimal* restoration based on a maximum probability criterion. For the Gibbs sampler, on the other hand, there is usually no unique[2] optimal restoration. In this case, convergence towards the final result would actually refer to convergence of the sample statistics towards the equilibrium distribution, which is given by the joint density of the unknown parameters.

The number of iterations required to attain convergence is strongly dependent on the proportion of missing data. For the examples discussed in this chapter the ML and EM algorithms needed about ten iterations to converge. As a rule of thumb the Gibbs sampler seems to require about twice as many iterations, if not more. The Gibbs sampler also requires significantly more computation than either the ML or EM algorithms. This is because of the need to compute the square roots of the covariance matrices to facilitate sampling of the multivariate Gaussian statistics. Overall, in terms of the length of CPU time needed, the Gibbs sampler seems to take four times as much effort to give the final result.

The examples discussed in this chapter provided quite a severe test for the algorithm. The gaps that were interpolated were extraordinarily long. This accounts for the large number of iterations required to achieve convergence. In more typical examples only five or so iterations are needed.

One major advantage of the Gibbs sampler over rival techniques is that it is a global optimization technique. One problem with the ML and EM techniques is that they can potentially get trapped by local extrema in the likelihood surface. This cannot possibly happen when using the Gibbs sampler.

[2]Exceptions include the sine wave described earlier.

6.9.3 *Modelling issues*

Interpolation of missing data using an autoregressive model has proven to be an extremely elegant application of the Gibbs sampler. The key to the ease of use of the Gibbs sampler is that the three conditional densities, for the missing data, autoregressive parameters and standard deviation are all easily simulated.

The autoregressive model which underlies all this work is an extremely flexible model to use. One extreme case that we have seen is the sine curve (poles on the unit circle and no excitation). At the other extreme, white i.i.d. Gaussian noise is also a special case of the AR model. However, despite its flexibility it is not always applicable. First, it assumes that the signal is stationary which certainly isn't true for music. At best we can assume local stationarity and use an AR model to model the signal over short intervals (<50ms or so). Second, it requires rather large model order when the signal is pseudoperiodic and noisy. An example of an ARMA process was presented which required eighty AR coefficients before the signal could be modelled satisfactorily.

7

Integration in Bayesian Data Analysis

The aim of this chapter is to demonstrate usage of the numerical methods developed in chapters 3 and 4 for the Bayesian analysis of data. The chapter begins with an example for which the Bayesian evidence (or rather the integrated likelihood) and marginal densities may be computed in closed form. These numerical techniques are then applied to a difficult problem in data analysis; namely, inferring the number of decaying exponentials and the values of the decays in real experimental data. The chapter concludes with examples of model selection by determining the appropriate noise model and signal model to use in a given data set.

7.1 Polynomial data

Let us analyse a simple test problem to investigate the accuracy of the techniques described in this book. The problem of interest is the fitting of polynomials to data such as that shown in figure 7.1.

A polynomial fit to data may be written in the form of the general linear model

$$\mathbf{d} = \mathbf{G}\,\mathbf{b} + \mathbf{e} \tag{7.1}$$

where the matrix $\mathbf{G}$ consists of elements $G_{ij} = x_i^{j-1}$, and the elements of the residual vector $\mathbf{e}$ are i.i.d. Gaussian random variates.

For simplicity we will assume that the prior densities for all the parameters in the model are very diffuse. The posterior density may therefore be considered as being the likelihood function multiplied by some scaling fac-

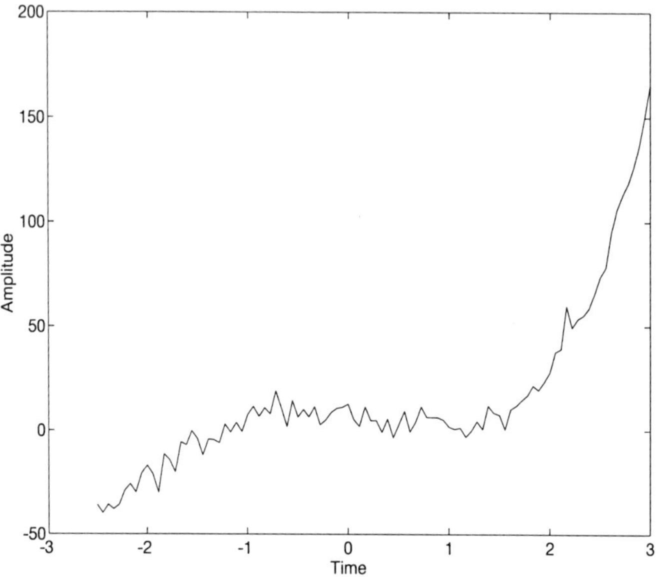

FIGURE 7.1. Fourth order polynomial data

tor that depends on the prior. In this case the likelihood may be integrated in closed form, so analytic expressions may be derived for the marginal density and evidence.

We shall integrate the likelihood over all the parameters, except the one of interest, to obtain any desired marginal density in closed form. The marginal density may be normalized quite easily by dividing by the integral of the likelihood over all the parameters in the model. The hypervolume under the likelihood function is proportional to the evidence, the constant of proportionality being the prior dependent scaling factor referred to above. In this section, we shall simply present results for the integrated likelihood, it being understood that a prior factor is also required for model selection. We shall show that the results of numerical integration agree with the theoretical value for the integrated likelihood given in equation 7.5 below.

Exact expressions for the integrated likelihood, as well as for the marginal densities of each of the parameters in the model are derived in appendix A. Exact expressions for the conditional densities for use in Gibbs sampling, and in numerical integration for computing marginal densities are also derived in appendix A.

7.1.1 Polynomial data

Consider the data set plotted in figure 7.1 which shows synthetic data of the form:

$$d_i = (t_i + 5.5)(t_i + 1.2)(t_i - 1.2)(t_i - 0.5) + e_i \tag{7.2}$$

where e_i is drawn from a zero mean i.i.d. Gaussian process with standard deviation $\sigma = 5$.

The aim is to calculate selected marginal densities and the evidence for the model.

7.1.2 Sampling the joint density

Three different Monte Carlo methods will be used to generate jointly distributed samples to compute the marginal density:

1. **Full Gibbs sampler**: Each parameter of the model is sampled in turn.

2. **Condensed Gibbs sampler**: The linear parameters are sampled altogether at once, and then the standard deviation is sampled.

3. **Hybrid Monte Carlo**: The energy function for the parameters of the model is simulated as a time dependent dynamical system.

In each case we can form the marginal density either by forming a histogram of the samples of the parameter of interest or by averaging the conditional density using the techniques developed in appendix E. Evidence may also be computed using reverse importance sampling as described in section E.4.

The Gibbs sampler

Two variants of the Gibbs sampler will be employed. The first variant which we term the *full Gibbs sampler* samples all the parameters of the model individually in turn. The linear coefficients are each sampled from a one dimensional conditional density as given by expression A.23. This is in a very simple Gaussian form. The standard deviation is sampled from the conditional density in expression A.29 which is in the form of an inverse chi density. Sample variates may be generated from an inverse chi density using the methods described in section 6.5.4.

In the second variant, which we term the *condensed Gibbs sampler*, the parameter space is partitioned into subspaces $\mathbf{b}$ and σ. The Gibbs sampler then has only two stages:

1. Draw a random Gaussian vector from the conditional density $p(\mathbf{b} \mid \sigma, \mathbf{d}, \mathbf{I})$.

2. Draw a sample of standard deviation from the conditional density $p(\sigma \mid \mathbf{b}, \mathbf{d}, \mathbf{I})$.

Step 1 involves drawing a random sample from a multivariate Gaussian distribution with mean $\left(\mathbf{G}^{\mathbf{T}}\mathbf{G}\right)^{-1}\mathbf{G}^{\mathbf{T}}\,\mathbf{d}$ and covariance matrix $\sigma^2\left(\mathbf{G}^{\mathbf{T}}\mathbf{G}\right)^{-1}$. This may be implemented using the methods described in appendix B.

Two thousand five hundred samples were drawn from the joint density using the full Gibbs sampler and the condensed Gibbs sampler. Figure 7.2 shows one thousand successive samples of the third coefficient (i.e. the x^2 term) of the polynomial. The trajectory followed by the samples is in the form of a random walk. The trajectory followed by the full Gibbs sampler is plotted as a solid curve, and that of the condensed Gibbs sampler plotted as a dotted grey curve. It is immediately apparent that successive samples from the full Gibbs sampler are highly correlated, whilst the samples from the condensed Gibbs sampler appear to be only weakly correlated, if at all. The huge difference in the random walks is due to the strong correlations between the linear coefficients[1]. Hence, the log likelihood surface is similar in form to that shown in figure 3.2. The major axis of the likelihood surface is over fifty times longer than the minor axis, for the data set shown in figure 7.1, so the "long narrow valley" effect is very pronounced. The presence of strong serial correlations can seriously degrade or bias an estimate of an integral, since the samples from a short run may not be truly representative of the density being simulated. This point is illustrated in the examples which follow.

Hybrid Monte Carlo

The Hybrid Monte Carlo algorithm requires the gradient of the energy function in order to simulate the dynamics of the system. The gradients required to implement the general linear model are presented in full in appendix C. For this application the required gradients are given by expression C.3 for the linear coefficients $\mathbf{b}$, and expression C.4 for the standard deviation σ.

The implementation of the Hybrid Monte Carlo algorithm is less problem specific than many other Markov chain approaches. However, two parameters that are of particular importance are the step size used in the leapfrog iteration, and the number of leapfrog steps to be taken during each transition. These two parameters must be tuned according to a given problem. Neal [82] analyses the behaviour of the leapfrog algorithm when simulating the dynamics of a multivariate Gaussian density. He shows that if the step size is greater than twice the minimum standard deviation of the density then the leapfrog discretization is unstable and will diverge.

---

[1]The correlations between the standard deviation and the linear parameters are very weak.

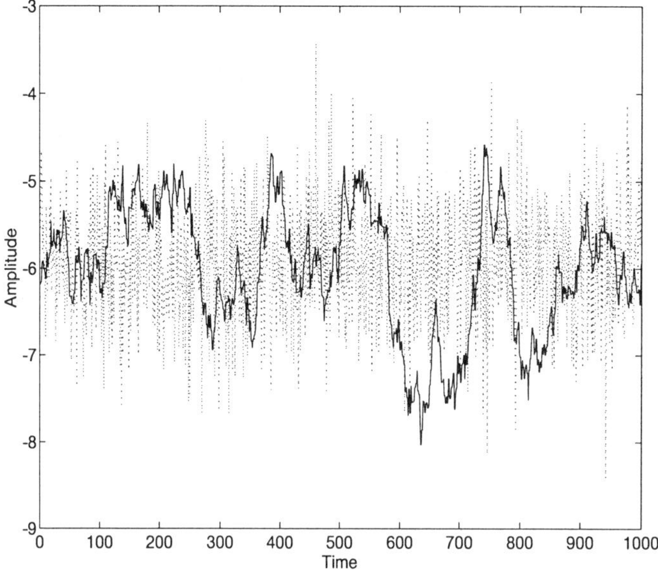

FIGURE 7.2. Trajectory followed by second order coefficient using Gibbs sampling: (solid) path followed when using full Gibbs sampler: (dotted) path followed when using condensed Gibbs sampler.

Based on this we can define a set of two heuristics for implementing the Hybrid Monte Carlo algorithm:

1. Choose the step size to be slightly less than twice the minimum standard deviation (i.e. $\epsilon < 2\sigma_{\text{min}}$).

2. Choose the number of steps to be equal to the ratio of the length of the major to the minor axis (i.e. $N \approx \sigma_{\text{max}}/\sigma_{\text{min}}$).

The second rule ensures that the trajectory can propagate the full length of the density along the major axis, so successive variates drawn from the Hybrid Monte Carlo algorithm are reasonably independent of one another.

The sizes of the required standard deviations can be obtained from a eigendecomposition of the full Hessian matrix. This will define a set of orthogonal axes, two of which correspond to the major and minor axes. The square root of the maximum eigenvalue of the Hessian matrix is the reciprocal of the minimum standard deviation.

$$\sigma_{\text{min}} = \frac{1}{\sqrt{\lambda_{\text{max}}}} \tag{7.3}$$

The number of steps can be set to the square root of the ratio of the maximum eigenvalue to the minimum eigenvalue.

$$N \approx \sqrt{\frac{\lambda_{\max}}{\lambda_{\min}}} \tag{7.4}$$

For this data set, sensible values of the step size and number of steps were found to be 0.01 and 50 respectively.

7.1.3 Approximate evidence

In all no less than *four* feasible approaches to determining model evidence are presented.

1. *Gaussian approximation*
2. *Non-iterative importance sampling*
3. *Reverse importance sampling*
4. *Thermodynamic integration*

Many variations of each of the above are possible. Table 7.1 gives the results of a number of different approaches to integrating the likelihood. The true value of the integral is known exactly from expression 7.5 below.

Integrated likelihood in closed form

The hypervolume enclosed under the likelihood function (i.e. the integrated likelihood) is given by the expression:

$$I = \frac{1}{2}\Gamma\left(\frac{N-M}{2}\right)\left[\pi\left(\mathbf{d}^{\mathbf{T}}\mathbf{d} - \mathbf{f}^{\mathbf{T}}\mathbf{f}\right)\right]^{-(N-M)/2}\left|\mathbf{G}^{\mathbf{T}}\mathbf{G}\right|^{-1/2} \tag{7.5}$$

Technique	Evidence
True	1.5224×10^{-129}
Gaussian approximation	0.8497×10^{-129}
Importance sampling (Random)	$1.53 \pm 0.97 \times 10^{-129}$
Importance sampling (Sobol)	$1.49 \pm 0.93 \times 10^{-129}$
Thermodynamic integration	$1.47 \pm 1.77 \times 10^{-129}$
Reverse importance sampling (HMC)	1.48×10^{-129}
Reverse importance sampling (CG)	1.48×10^{-129}
Reverse importance sampling (FG)	0.99×10^{-129}

TABLE 7.1. Evidence for polynomial data

where

$$\mathbf{f} = \mathbf{G}\left(\mathbf{G}^{\mathrm{T}}\mathbf{G}\right)^{-1}\mathbf{G}^{\mathrm{T}}\mathbf{d}$$

Gaussian approximation

The Gaussian approximation used here is derived from expression 3.15. The Hessian matrix is available in closed form, as is the mode of the likelihood function.

Importance sampling

The approach used here is to construct an importance sampling density as described in section 3.8.

One thousand independent random samples were drawn from a hyberbolic Cauchy density. These random samples were then transformed in order to match the importance sampling density to the posterior density. The Hessian matrix and the mode of the posterior density are required to do this. Both of these quantities are known in closed form [2].

Figure 7.3 shows the values of the ratio $f(\mathbf{x}_i)/g(\mathbf{x}_i)$, where $f(\mathbf{x})$ is the likelihood and where the samples $\mathbf{x}_i$ are generated from the importance sampling density $g(\mathbf{x})$. Each value of this ratio is, in effect, an unbiased estimate of the integral, and may be termed a *sample integral.* Except for the presence of one or two outliers the samples are loosely clustered about the mean.

The same analysis was repeated using quasi-random samples drawn from a Sobol' sequence. The resultant integral was not significantly better for this example, having only a marginally smaller standard deviation. The Sobol' sequence should give a more accurate result than random sampling if more samples are taken.

The hyperbolic Cauchy density is well suited to sampling Gaussian parameters such as the model amplitudes (i.e. the polynomial coefficients). However, it is not suited to sampling skewed densities such as that of the standard deviation which has an inverse chi density. This would perhaps explain the presence of the outliers. This is not a serious problem here but if, in applying the method to another data set, it were observed that the problem became severe then one could easily improve matters by choosing an appropriate reparameterization.

Thermodynamic integration

The method of thermodynamic integration was described in section 4.9.1.

In this section we use thermodynamic integration to estimate the free energy difference between the likelihood function and its Gaussian approx-

[2] These were computed earlier whilst determining the Gaussian approximation.

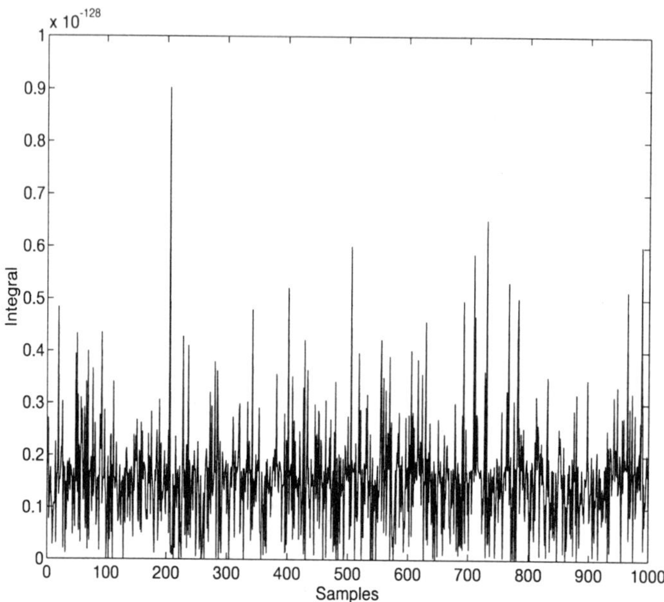

FIGURE 7.3. Trial integrals obtained using importance sampling while computing evidence

imation. Since the integral for the Gaussian approximation is known exactly, we therefore have a means of estimating the *absolute* free energy of the likelihood function.

Ninety nine intermediate systems were defined, with α in equation 4.44 ranging from 0 to 1 in uniform steps of 0.01. The position was updated at each stage by means of a Metropolis step. The random starting positions were sampled from the Gaussian approximation to the likelihood.

Figure 7.4 shows one thousand trial integrals obtained in this way. The mean of these trial integrals is given in table 7.1.

Reverse importance sampling

The principle of reverse importance sampling is described in appendix E.4, the essential idea being that random variates drawn from the likelihood are used to integrate a function, the normalization constant of which is actually known, thus giving an estimate for the normalization constant for the likelihood.

Table 7.1 gives the results of integrating a Gaussian approximation to the likelihood using one thousand successive random samples drawn from

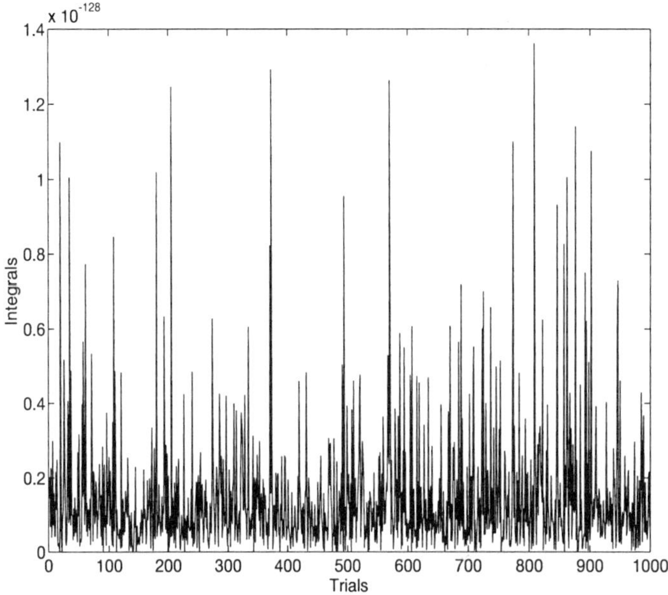

FIGURE 7.4. Thermodynamic integrals computed while calculating evidence

three different sources of random variates. The three sources used were Hybrid Monte Carlo (HMC) and two different versions of the Gibbs sampler; the condensed Gibbs sampler (CG) and the full Gibbs sampler (FG). The estimate obtained using the full Gibbs sampler is poor because of strong correlations between successive samples.

The mean and covariance matrix of the Gaussian approximation were estimated using expressions E.11 and E.12 from the sample realization, which is the same sample realization used to compute the integral. This may have introduced some bias into the result, but it does not seem to have been significant.

Remarks

It is interesting to note that the Gaussian approximation gave the worst performance[3] of all the techniques applied here, since it is probably the

[3]For some data sets the Gaussian approximation may be incorrect to two orders of magnitude! The accuracy tends to diminish with increasing dimensionality.

most common approach [13, 74] to approximating evidence for model selection.

Reverse importance sampling performed quite well. This is probably due to the closeness of fit between the likelihood and its Gaussian approximation. Importance sampling gave particularly good results, considering how few samples were used.

Thermodynamic integration performed far better than the error bars would suggest. The estimates are drawn from an impulsive random process with mean value equal to the desired integral. Because the sample estimates are impulsive, the standard deviation of the process is not an appropriate measure of the error. Thermodynamic integration has the great advantage that it is readily applicable to many situations, because once enough intermediate systems are introduced the free energy difference may be computed regardless of the differences between the two systems being compared.

7.1.4 Approximate marginal densities

In this section, we shall attempt to evaluate the marginal densities of the parameters of the polynomial model using numerical methods.

Two classes of approach are used:

1. *Asymptotic approximations*

2. *Monte Carlo integration*

Three different asymptotic approximations are investigated. These are the Gaussian approximation described in section 3.1.6, as well as the Laplace and profile approximations of section 3.5.

Monte Carlo methods for marginalization can be divided into three categories. First, one may simulate jointly distributed variates by whatever means are convenient (e.g. using the Gibbs sampler, Metropolis algorithm, Hybrid Monte Carlo, SIR etc.) and estimate the marginal density from the samples of a component of the random variates. Second, one may make alternative use of the jointly distributed variates by averaging the conditional density (as described in section 3.9.4 and in more detail in appendix E.1). We also consider approaches that do not require jointly distributed variates. These are the dummy variable method of section 3.9.3 as well as a modified version of the SIR algorithm described in section 3.9.2.

Marginalization using asymptotic approximations

Figure 7.5 shows the marginal density for the second parameter (i.e. the coefficient in x^1) of the polynomial as a solid curve. The theoretical density is in the form of a student-t distribution given by expression A.18.

The corresponding Gaussian approximation is plotted as a dashed curve and the profile approximation is plotted as a dotted curve. The Gaussian

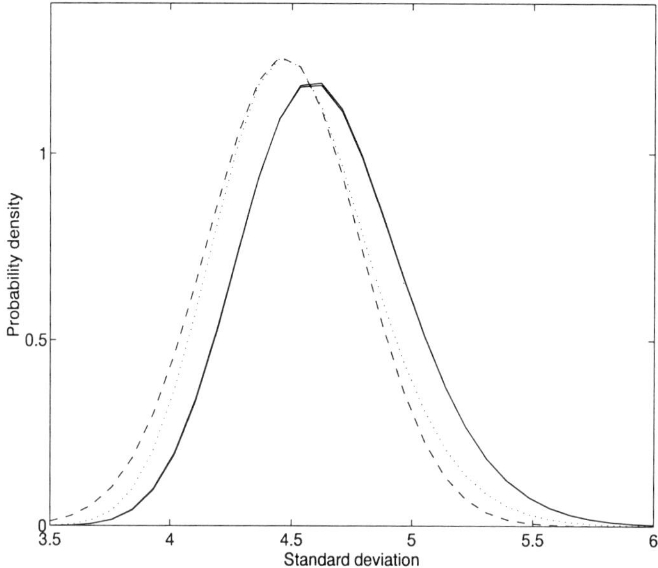

FIGURE 7.5. Asymptotic approximations to the marginal density for linear coefficient: (dashed) Gaussian approximation; (dotted) profile approximation: (solid) Laplace approximation and true marginal density.

approximation overestimates in the shoulders of the distribution and underestimates at the mode. This is the opposite of the behaviour of the profile approximation.

The Laplace approximation is also shown but cannot be distinguished from the true marginal density.

Figure 7.6 shows the marginal density for the standard deviation. The Gaussian approximation is plotted as a dashed curve and the profile approximation as a dotted curve. The true marginal density and the Laplace approximation are plotted as solid curves.

In this case, the Gaussian and profile approximations are quite clearly incorrect. The shape of the density is not skewed enough and the estimate, as given by the position of the mode, is biased. The Laplace approximation and the true marginal density are almost indistinguishable.

In this particular example, the conditional maxima and derivatives (such as the Hessian matrix) needed for asymptotic approximations could be evaluated in closed form. For example, the conditional maximum for a linear coefficient that was used above to approximate marginal densities, is given by the mode of expression A.18. However, exact expressions are not nor-

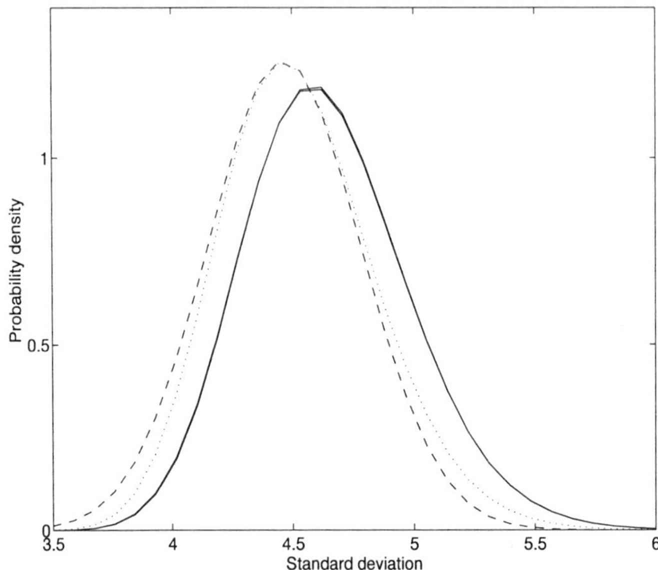

FIGURE 7.6. Asymptotic approximations to marginal density for standard deviation:(dashed) Gaussian approximation; (dotted) profile approximation; (solid) Laplace approximation and true marginal density.

mally available and for most problems conditional maxima and derivatives have to be computed numerically.

Marginalization using jointly distributed samples

Figure 7.7 shows the true marginal density and three estimates of the marginal density for the second linear coefficient. The three estimates were computed by averaging the conditional density over three different sources of jointly distributed variates: namely, the Hybrid Monte Carlo (HMC) method, the full Gibbs (FG) sampler and the condensed Gibbs (CG) sampler. One thousand successive variates from the random sequence were used in each case.

The HMC estimate of the marginal density is plotted as a dotted curve. The CG estimate and FG estimate of the marginal density are plotted as dashed and dot-dashed curves respectively. The agreement between the HMC and CG estimates and the true marginal density (solid curve) is quite good. The FG sampler, to put it quite succinctly, is a disaster.

The reason for the very poor performance of the FG sampler is the large correlations between successive samples. This does not mean that the FG

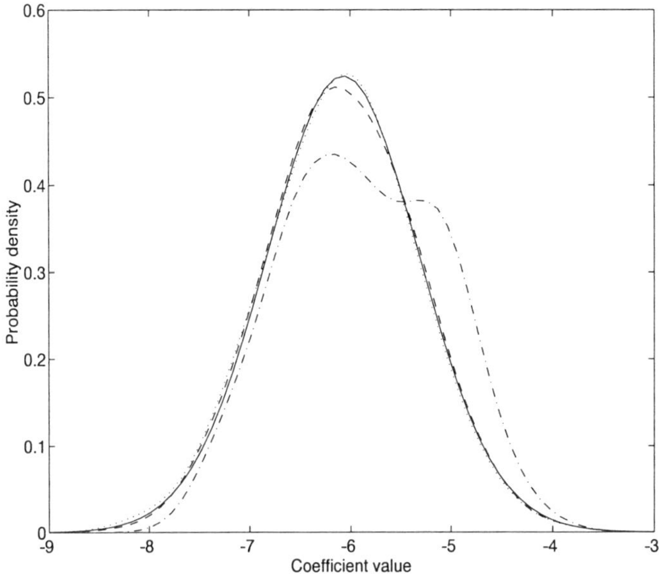

FIGURE 7.7. Monte Carlo integrals for the marginal density of the first order linear coefficient: (dotted) Hybrid Monte Carlo; (dot-dashed) Full Gibbs sampler; (dashed) Condensed Gibbs sampler; (solid) True.

sampler may not be used for integration, but that one should be aware of the problem with correlations, and thus generate a very large number of samples and from these extract samples that are widely spaced.

Figure 7.8 shows the true marginal density and three estimates of the marginal density for the standard deviation. As before, the three estimates were computed by averaging the conditional density over one thousand successive samples drawn from the HMC algorithm, the condensed Gibbs sampler and the full Gibbs sampler.

As in the previous case, the estimate given by the HMC algorithm (dotted curve) is the closest to the true marginal density. The CG sampler estimate (dashed curve) and the FG sampler estimate (dot-dashed curve) are also in close agreement with the true marginal density (solid). Not surprisingly, the CG estimate is better.

Figure 7.9 and figure 7.10 show the marginal density for the standard deviation obtained using the HMC algorithm and the CG algorithms respectively. Both bar graphs were obtained by forming a histogram of samples of the standard deviation component extracted from two thousand five hundred successive jointly distributed random vectors. The true marginal

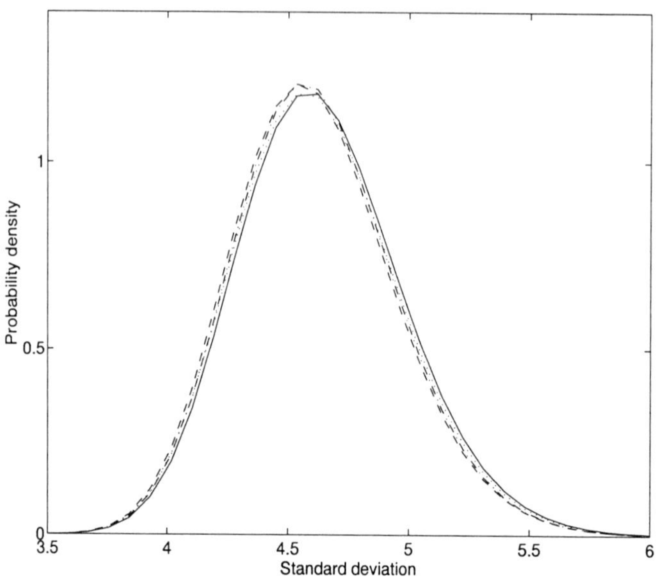

FIGURE 7.8. Monte Carlo integrals for the marginal density of the standard deviation: (dotted) Hybrid Monte Carlo; (dot-dashed) Full Gibbs Sampler; (dashed) Condensed Gibbs Sampler; (solid) True.

density is plotted as a dashed curve in each case.

Marginalization using recycled evidence variates

In section 3.9.3 we expressed a desire to reuse as much as possible of the computation carried out while evaluating the evidence when computing marginal densities. One approach that can be used to achieve this is the SIR algorithm described in section 3.9.2. Pseudo-jointly distributed random samples can be generated using weighted resampling and an estimate of the marginal density can be formed by averaging the conditional density. Alternatively, if the conditional density is not available, one can simply form histograms. Here we describe a modified SIR algorithm that is more efficient than either approach.

Let us express the marginal density as:

$$f(a) = \int f(a \mid b) f(a', b) \, da' \, db \tag{7.6}$$

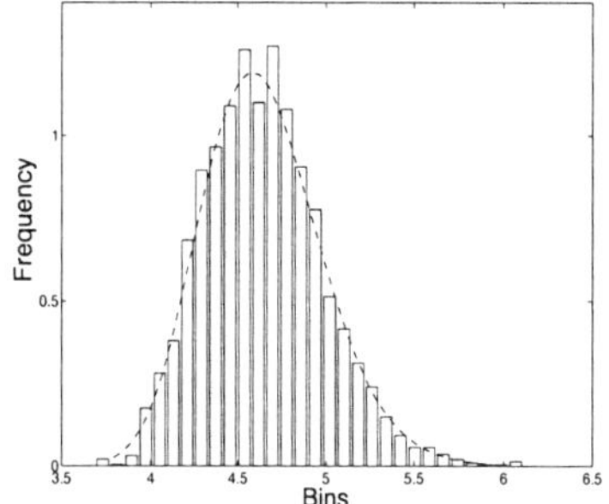

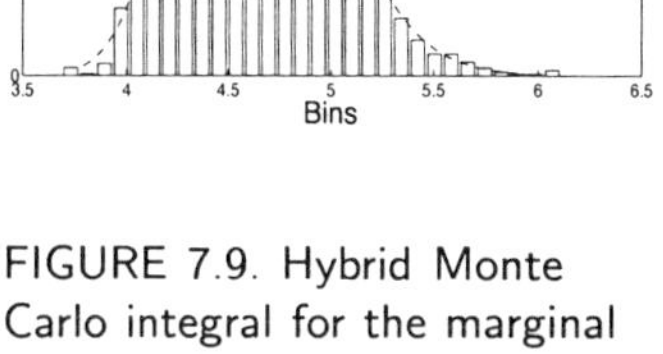

FIGURE 7.9. Hybrid Monte Carlo integral for the marginal density of the standard deviation – histogram (2500 samples)

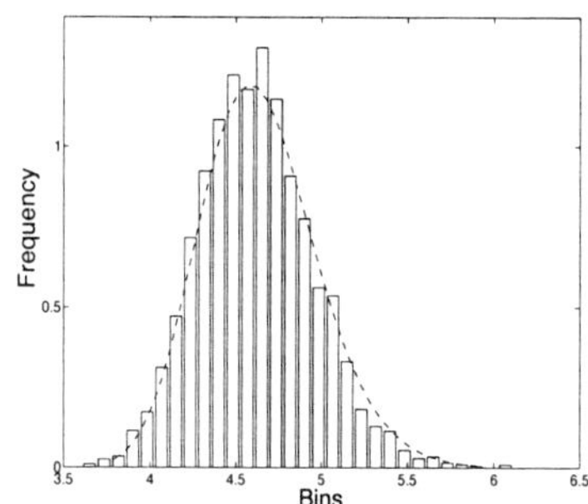

FIGURE 7.10. Gibbs sampler integral for the marginal density of the standard deviation – histogram (2500 samples)

Also, let us suppose that samples (a'_i, b_i) were generated from a density $g(a', b)$ in the course of computing the evidence. We may write:

$$f(a) = \int f(a \mid b) \frac{f(a', b)}{g(a', b)} g(a', b)\, da'\, db \tag{7.7}$$

which may be estimated using Monte Carlo integration as the following weighted sum:

$$f(a) \approx \frac{1}{N} \sum_{i=1}^{N} \frac{f(a'_i, b_i)}{g(a'_i, b_i)} f(a \mid b_i) \tag{7.8}$$

Note that the weights $f(a'_i, b_i)/g(a'_i, b_i)$ are each sample estimates of the evidence. The only computation required is to evaluate the conditional densities $f(a \mid b_i)$.

The dummy variable method (DVM) is different in that it is the joint density, rather than the conditional density, that is used. From expression 3.98 the marginal density is given by the expression:

$$f(a) \approx \frac{1}{N} \sum_{i=1}^{N} \frac{h(a'_i \mid b_i)}{g(a'_i, b_i)} f(a, b_i) \tag{7.9}$$

The conditional density $h(a' \mid b)$ can, in theory, be any pdf that is normalized over the parameter space. In this case, we shall employ a Gaussian approximation to the conditional density, where the mean and covariance matrix are estimated by maximization and numerical differentiation of the log density.

Figure 7.11 shows the marginal density for the standard deviation. The true marginal density is plotted as a solid curve. The SIR and DVM estimates of the marginal density are shown as a dashed curve and a dotted

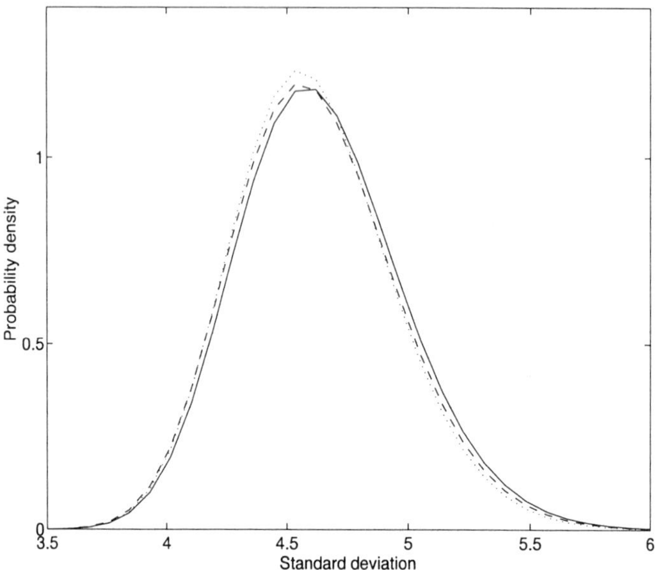

FIGURE 7.11. Monte Carlo integrals for the marginal density of the standard deviation; (dotted) Dummy variable; (dashed) SIR; (solid) True.

curve respectively. Both these estimates of the marginal density were plotted using the one thousand samples used to compute the sample estimates of the evidence in figure 7.3.

The estimate obtained using the SIR approach is more accurate than that obtained using the DVM approach. The most likely reason for this is that SIR uses just one importance sampler (i.e. weighting function), whereas DVM uses two.

7.1.5 *Conclusion*

This section reviewed the main ideas and techniques behind many of the numerical methods described in chapter 3 and chapter 4.

The performance of these methods was, for the most part, very good in that good approximations to the evidence (or integrated likelihood) and marginal densities were achieved using a reasonable amount of computation. Where the results were poor, such as in the case of the full Gibbs sampler, satisfactory reasons were given for the failure of the technique.

The polynomial example, for all its importance, was something of a toy problem because all the required integrals and marginal densities could be evaluated in closed form. In addition, the conditional maxima and derivatives needed to implement some of the numerical integration techniques could also be evaluated in closed form. This was to our advantage because it gave the opportunity to compare the results obtained using each method against the correct expressions. At the same time, most of the difficulties that can be expected of a less tractable problem occurred here. The fitting of polynomials to data is a rather ill-conditioned problem, which leads to very strong correlations between the individual parameters. In addition, non-Gaussian behaviour was observed in the case of one of the parameters (i.e. the standard deviation). Strong posterior correlations, as we have seen, slow the convergence of the full Gibbs sampler because of the "long narrow valley" effect. It also slows down the Hybrid Monte Carlo algorithm since more leapfrog steps are required per sample variate. Non-Gaussian behaviour in the parameter densities makes importance sampling more difficult. Both of these problems can be overcome by means of suitable transformations of the system parameters.

7.2 Decay problem

Numerous scientific and engineering applications involve fitting sums of decaying exponentials to experimental data.

$$d_i = \sum_{i=1}^{N} A_j \exp(\lambda_i t_i) + e_i \tag{7.10}$$

The amplitudes A_j and the decays λ_j must be estimated in some way. This is the classic problem described by Lanczos [67]. This is a very ill-posed problem, as observed by Lanczos [67], since slight fluctuations in the observed data can result in very large fluctuations in the estimated parameters. The ill-conditioning of this problem makes it difficult to solve satisfactorily. Methods designed just for estimating the parameters of decaying exponentials have been developed, starting with Prony [96]. Good introductions to the solution of the problem are given in the books by Lanczos [67] and Anderson [5]. Most techniques work well providing the decay rates are well separated and there is a large number of data and a high SNR. Common approaches to the problem include the exponential curve peeling method of Jacquez [54] and the singular value decomposition as described by Anderson [5].

In this section we use a fully Bayesian approach to investigate this problem. The number of component exponentials will be determined by comparing model evidence for different orders of the signal model. The decay parameters will be estimated using marginal densities.

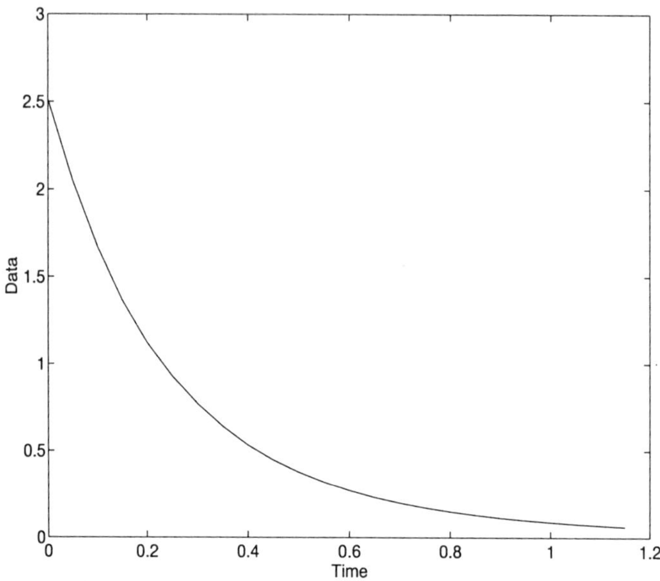

FIGURE 7.12. Lanczos decay data

7.2.1 The Lanczos problem

Lanczos [67] considers the following function:

$$f(t) = 0.0951 \exp(-t) + 0.8607 \exp(-3t) + 1.5576 \exp(-5t) \tag{7.11}$$

Table 7.2 presents expression 7.11 evaluated to four decimal places, with 24 data points where t ranges from zero up to 1.15 in steps of 0.05. This means that the data is corrupted by quantization noise. It is concluded in the book of Lanczos that the problem is so ill-conditioned, that even with the very low noise levels in the data, it is quite impossible to distinguish between the "true" fit and that of the function:

$$\tilde{f}(t) = 2.202 \exp(-4.45t) + 0.305 \exp(-1.85t) \tag{7.12}$$

within the given range of t. The function $f(t)$ is plotted in figure 7.12.

We shall examine this statement now, using Bayesian parameter estimation to determine the decays, and Bayesian model selection to determine the order of the model. We shall confine our interest to the decay rates so

Truncated Lanczos Data					
Time	Data	Time	Data	Time	Data
0.0000	2.5134	0.4000	0.5337	0.8000	0.1493
0.0500	2.0443	0.4500	0.4479	0.8500	0.1300
0.1000	1.6684	0.5000	0.3775	0.9000	0.1138
0.1500	1.3664	0.5500	0.3197	0.9500	0.1000
0.2000	1.1232	0.6000	0.2720	1.0000	0.0883
0.2500	0.9268	0.6500	0.2324	1.0500	0.0783
0.3000	0.7679	0.7000	0.1996	1.1000	0.0697
0.3500	0.6388	0.7500	0.1722	1.1500	0.0623

TABLE 7.2. The Lanczos data

we begin by integrating out[4] the amplitudes $\mathbf{b}$ and the standard deviation σ as nuisance parameters.

First, we shall locate the mode of the posterior density for the two and three decay models. Second, we shall integrate to determine the marginal densities of the decay rates in each case. Both objectives are very easily achieved using the HMC algorithm, with optimization being carried out using simulated annealing as described in section 4.7.2, and integration by simulating the dynamics at unit temperature and forming histograms. Finally, we shall determine the evidence for the data for a set of exponential models of order 1-4 inclusive.

Optimization

In this problem, the energy function is multimodal but each mode is identically equivalent to all other modes. To see this one needs only consider equation 7.10. A fit obtained by using the parameter set $(A_1, \lambda_1, A_2, \lambda_2)$ is completely identical to the fit obtained using $(A_2, \lambda_2, A_1, \lambda_1)$. If one ignores the effects of permuting the parameters then one can state that the energy function is unimodal.

There is therefore no real need to initialize the temperature to a high value. In addition, a moderately low value for the "memory factor" α in equation 4.31 is appropriate since a large value will result in unnecessarily slow convergence of the algorithm. For this problem, we set $\alpha = 0.1$.

The step size and the number of steps were set (after a little experimentation) at $\epsilon = 0.001$ and $N = 50$ respectively. The start point for the HMC algorithm was arbitrarily set at $\alpha_1 = -0.5$ and $\alpha_2 = -1$. Figure 7.13 shows the subsequent evolution of the value of the energy function in time. Figure 7.14 shows the trajectory followed by the decay parameter vector in time. The initial position for the decays is in the top right hand side

[4]We shall assume Gaussian noise statistics for ease of analysis.

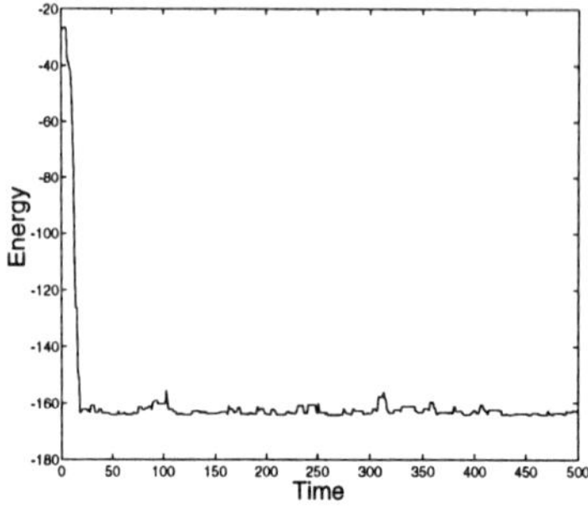

FIGURE 7.13. Hybrid Monte Carlo optimization: Energy against time

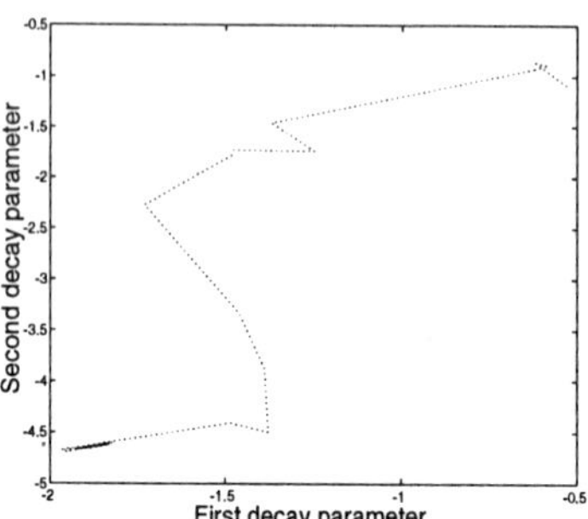

FIGURE 7.14. Trajectory for two decay model

of the figure at position (−1, −0.5). After less than fifteen transitions the trajectory has converged to the mode in the bottom left hand corner and thus towards equilibrium. The strong line of direction in the trajectory is obvious. The posterior mode is just visible as a dark streak on the bottom left hand corner of figure 7.14. It is interesting that out of one thousand samples plotted, 5% lie outside the posterior mode from the initial "burn in" or period of optimization. The remaining 95% lie within the body of the posterior density.

Marginal densities

The marginal densities were estimated by simulating the joint posterior density using Hybrid Monte Carlo and forming histograms.

Figures 7.15 and 7.16 show the marginal densities $p(\lambda_1 \mid \mathbf{d}, \mathbf{I})$ and $p(\lambda_2 \mid \mathbf{d}, \mathbf{I})$. The estimated decays broadly correspond to the values used by Lanczos to fit the data.

Figures 7.17, 7.18 and 7.19 show the marginal densities for each of the decay rates in the three exponential model. The three estimates agree well with the "true" values used when generating the data.

The accuracy of the estimates, or more precisely, one's confidence in the inferred values for the decays, is reflected in the width of the densities. Not surprisingly, the larger is the rate of decay, the larger is the corresponding width of the marginal density. This is because if the decay rate is larger then that particular component of the signal will disappear more quickly. This means that the larger decay rates are effectively being determined by a smaller number of data.

The marginal densities of the decays in the three decay model are quite skewed. This is particularly apparent for the fastest decay rate. The non-

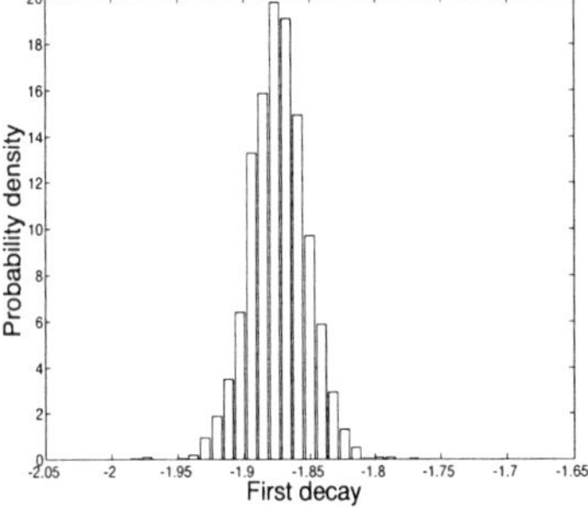

FIGURE 7.15. Two decay model: Marginal density for first decay

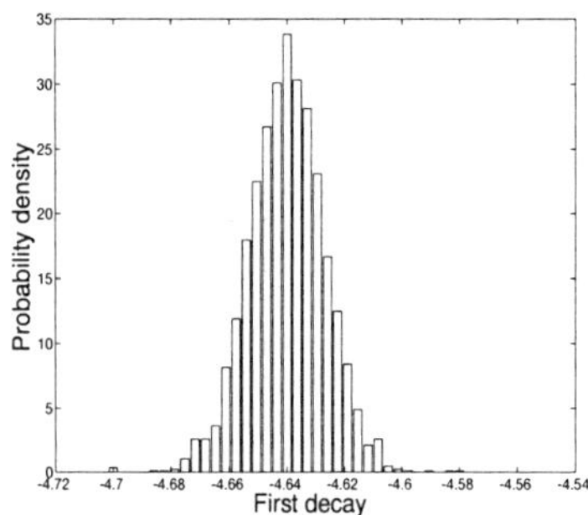

FIGURE 7.16. Two decay model: Marginal density for second decay

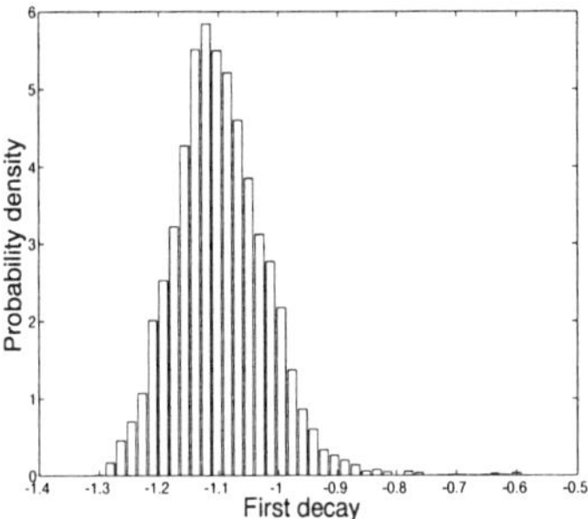

FIGURE 7.17. Three decay model: Marginal density for first decay

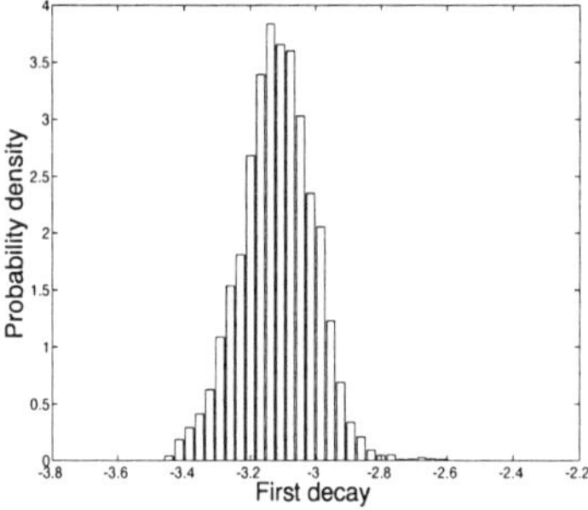

FIGURE 7.18. Three decay model: Marginal density for second decay

Gaussianity of the marginal densities suggests that complex interactions are taking place between the parameters of the model in fitting the data.

Figure 7.20 is a plot of the samples of the smallest decay against the corresponding samples of the largest decay. The result is termed a *probability cloud.* The non-Gaussianity of the joint density is immediately apparent. The interactions between the parameters gives rise to a rather surprising, and very interesting, crescent shape. This crescent shape is often observed in other cases as well, and can be far more pronounced than in the example shown here. One example of when it may occur is when one is attempting to resolve between two closely spaced exponentials.

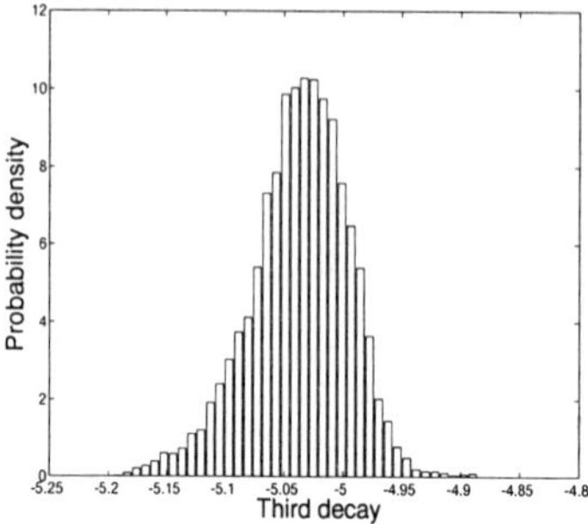

FIGURE 7.19. Three decay model: Marginal density for third decay

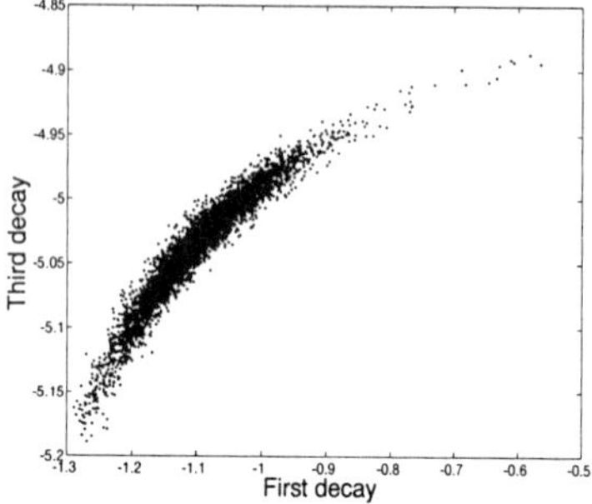

FIGURE 7.20. Three decay model: Probability cloud for two parameters

Model selection

The real problem that Lanczos had with analyzing the data was not in estimating the parameters, but in estimating the correct model order.

In this section three different methods are applied to estimate the model order.

Quadrature: The dimensionality of the integrals are quite low here so numerical quadrature is a feasible approach. A 23 point Simpson's rule (i.e. uniform sampling using weights $\frac{1}{3}\{1\ 4\ 2\ 4\ 2\ \cdots\ 2\ 4\ 1\}$ was used to compute the integrals in table 7.3. The disadvantage of using this approach is that the limits on the integral must be finite, but this is probably not a serious problem providing the bounds of the integrals lie well within the tails of the joint posterior density. The bounds on the multiple integral computed using Simpson's rule are given in table 7.4.

Gaussian approximation: An integrated normal approximation to the likelihood was obtained using a BFGS numerical optimization procedure. The starting points for the numerical optimization had to be obtained from the Hybrid Monte Carlo algorithm samples obtained above. Arbitrary starting points led to numerical instability or a sub-optimal minimum in the energy function. The latter could not have been caused by local minima in the cost function because there are none. It is more likely to be a result of some unusual local condition, such as a plateau on the energy function surface or numerical instability, triggering the stopping criteria for the algorithm.

Thermodynamic integration: The Gaussian approximation may be used as a comparison system for the thermodynamic integral. In this way

Model order	Evidence		
	Quadrature	Gaussian approx.	Thermo. integration
1	7.74×10^{20}	7.47×10^{20}	7.71×10^{20}
2	1.06×10^{57}	9.68×10^{56}	1.09×10^{57}
3	8.03×10^{79}	1.61×10^{78}	3.05×10^{78}
4	8.86×10^{77}	4.27×10^{69}	1.02×10^{72}

TABLE 7.3. Evidence for decay data

Model order	Parameter	Limits on integral	
		Lower bound	Upper bound
1	α_1	-4.2	-3.5
2	α_1	-4.7	-4.6
	α_2	-2.0	-1.75
3	α_1	-1.4	-0.7
	α_2	-3.6	-2.6
	α_3	-5.25	-4.85
4	α_1	-10	α_2
	α_2	-10	α_3
	α_3	-10	α_4
	α_4	-10	-1.5

TABLE 7.4. Bounds on multiple integrals computed using Simpson's rule

the free energy difference may be computed which leads to an estimate of the absolute evidence. Ninety nine intermediate systems were defined with the mixture parameter α in equation 4.44 varying uniformly from 0 to 1 in steps of 0.01.

All the integration methods agree that the model order, as given by the model order with the highest relative evidence, is three. This agrees exactly with the model order in equation 7.11 that was used for generating the data.

The agreement between the results obtained using the three different techniques is good for low orders of model but becomes poor for the highest order considered. The results obtained for the four decay model especially require discussion.

The quadrature integral is quite trustworthy for low model order, but is unreliable for the higher orders. One contributory factor is systematic errors in the product rule. Simpson's rule is a third order interpolatory formula [28, 124] which means that it attempts to fit the samples from the integrand as a piecewise cubic curve. Indeed there is a good case for using the trapezoidal rule instead! A 9 point Newton-Cotes formula was also tried but this gave very poor results (i.e. sometimes negative!). This is because the 9th order polynomial used to interpolate the samples (that is implicit in this formula) was obviously overfitting.

Gauss-Hermite quadrature was also an option but this proved rather cumbersome to apply, because of the need for conditional maximization and numerical differentiation to compute local Gaussian approximations to the density in order to place the samples optimally. The exact details of implementation are similar to those of the Laplace approximation.

The number of computations required to evaluate each integral for model orders $\{1, 2, 3, 4\}$ were $\{23, 529, 12167, 279841\}$ respectively. The last integral (four dimensional) took just under half an hour on a 60MHz Pentium PC. A five dimensional integral would require half a day!

Gaussian quadrature, if anything, seemed to underestimate the evidence. This is especially true for model order equal to four. The fourth order result quoted in table 7.3 is probably in error. This is most probably due to the finite step size used in the numerical differentiation. The value of a determinant of a matrix can be very sensitive to small changes in the individual elements. It may be that just one of the 16 elements of the Hessian matrix may be slightly wrong, causing a catastrophic change in the value of the integral.

Thermodynamic integration is the most believable result quoted here. There are a number of reasons for this. During the calculation one hundred trial integrals were computed, each of which should have given an independent unbiased estimate of the free energy difference. It is possible to obtain an estimate of the error from the standard deviation of the trial integrals. After ignoring obvious outliers, this indicated an error of approximately 50%.

A multivariate Gaussian with a diagonal covariance matrix was also tried as a comparison system. The covariance matrix was calculated form the range in table 7.4, the width of the range on each parameter in table 7.4 being equal to four standard deviations. The results obtained were in agreement with those of table 7.3.

The amount of time needed to compute the four dimensional thermodynamic integral was not much greater than the time needed to compute the one dimensional integral (i.e. approximately six times more). No more than a few minutes were required to carry out one hundred integrals.

To conclude, it would seem from this example that thermodynamic integration is the only viable method of the three for computing evidence for difficult models.

7.2.2 Biomedical data

Consider the data plotted in figure 7.21 which shows the results of a biomedical experiment. The data consists of four hundred data points. The first

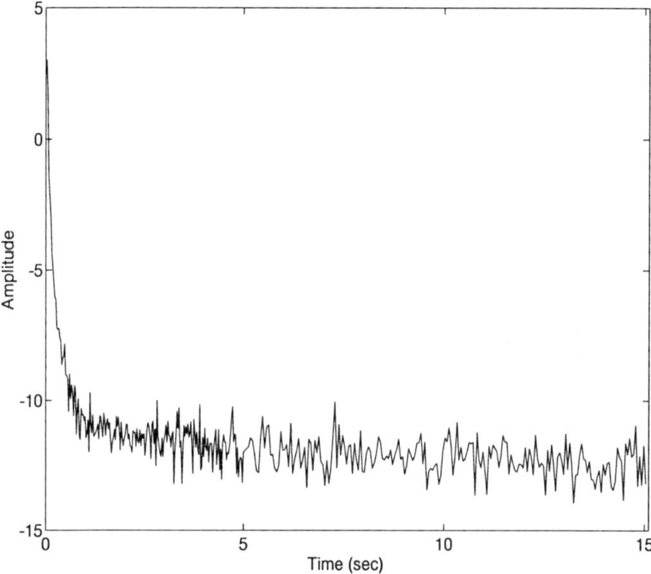

FIGURE 7.21. Biomedical data

two hundred data points were sampled at a time intervals of $25ms$, and the remaining two hundred data points every $76ms$.

The data are the result of a stopped flow experiment [7] in which two solutions are mixed very rapidly by injecting them into a common observation chamber. The reaction is measured by observing the spectroscopic properties of the compounds involved. The reaction of interest is the reformation of the structure of a protein as the pH is increased.

The data is believed to be generated by a model of the following form:

$$f(t_i) = A_1 \exp(\lambda_1 t_i) + A_2 \exp(\lambda_2 t_i) + c + e_i \tag{7.13}$$

where the residuals are sampled from an i.i.d. Gaussian noise process $N(0, \sigma^2)$.

The aim of the experiment is to determine *all* the parameters, namely the decay rates and the amplitudes, as well as the baseline and the standard deviation. The six dimensional parameter vector is ordered as follows: $\{\lambda_1, \lambda_2, c, A_1, A_2, \sigma\}$.

No parameters are integrated out in this problem. Therefore the objective function used to optimize the values of the parameters is the joint density.

Uniform prior densities are assumed for the parameters. The joint density is therefore taken to be proportional to the likelihood function.

Analysis

The data was analysed using the HMC algorithm. The gradients of the log likelihood function that are required to implement the HMC algorithm are given in section C.1. The starting point for the algorithm was arbitrarily chosen to be the point $\{-1, -1, -12, 1, 1, 1\}$. The step size ϵ was set at 0.01 and the number of leapfrog steps set at 50. The autoregressive parameter α in equation 4.31 was set at 0.5.

The trajectory followed by the algorithm for the two decay parameters is shown in figure 7.22. The corresponding values of the energy function are plotted in figure 7.23. It is apparent that the mode of the distribution was located quite quickly, but not as quickly as in the Lanczos example in section 7.2.1. This is probably due to the higher value of α.

The posterior density was sampled twenty thousand times in total using the HMC sampler. The marginal densities for each of the six parameters is shown in the following figures. Figures 7.24 and 7.25 show the marginal densities $p(\lambda_1 \mid \mathbf{d}, \mathbf{I})$ and $p(\lambda_2 \mid \mathbf{d}, \mathbf{I})$ respectively. As one might expect, the slower decay rate is well determined by the data while the value for the faster decay rate is not so well determined.

The marginal densities for the amplitudes $p(A_1 \mid \mathbf{d}, \mathbf{I})$ and $p(A_2 \mid \mathbf{d}, \mathbf{I})$ are shown in figures 7.26 and 7.27 respectively. The marginal density for the standard deviation $p(\sigma \mid \mathbf{d}, \mathbf{I})$ and the baseline $p(c \mid \mathbf{d}, \mathbf{I})$ are shown in figures 7.28 and 7.29.

There is an interesting feature in the marginal density for the baseline offset that is difficult to explain. In addition to the dominant mode at $c \approx -12.5$ there also appears to be a second mode at $c \approx -13.1$.

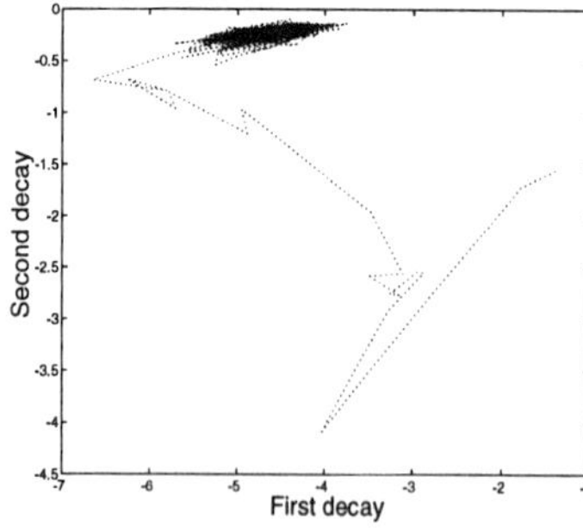

FIGURE 7.22. Two decay model: Trajectory followed by HMC algorithm

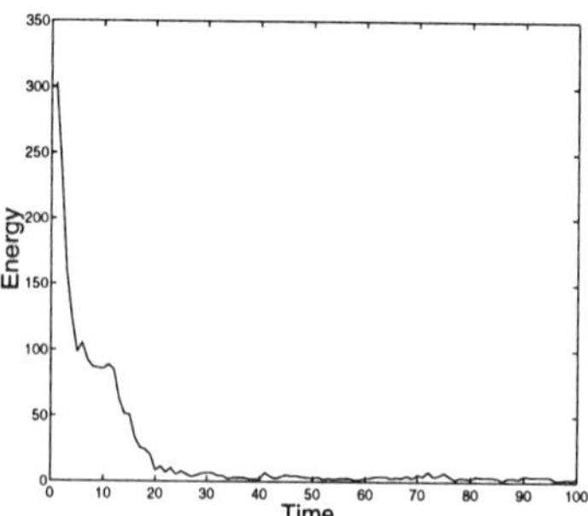

FIGURE 7.23. Two decay model: The energy optimization curve

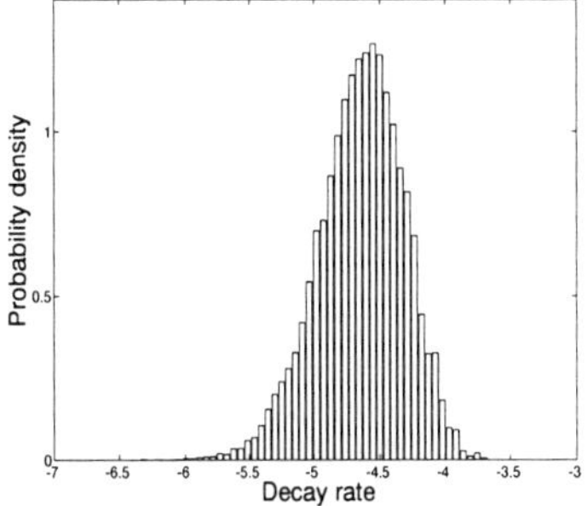

FIGURE 7.24. Two decay model: Fast decay rate

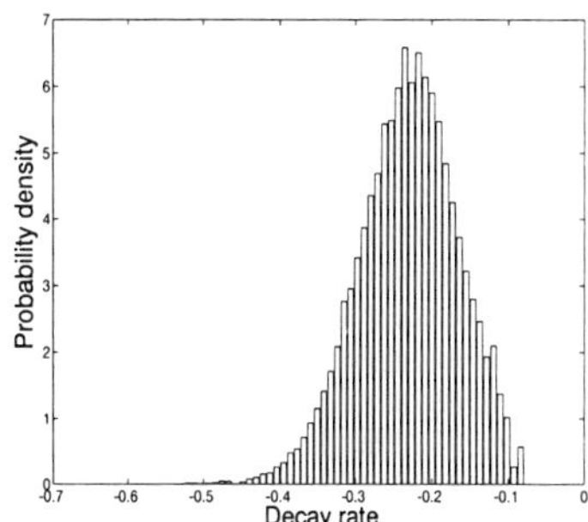

FIGURE 7.25. Two decay model: Slow decay rate

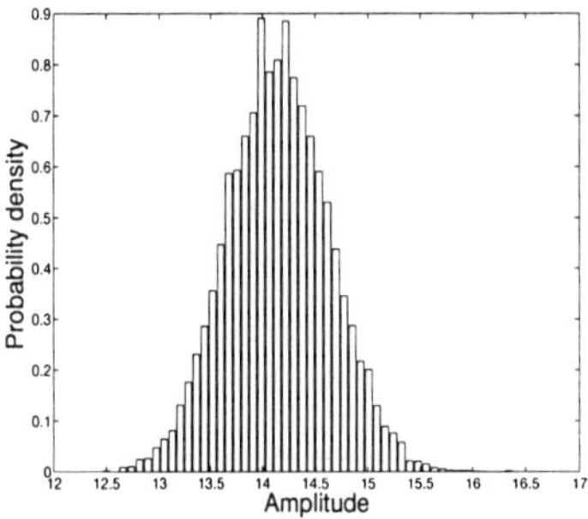

FIGURE 7.26. Two decay model: first amplitude

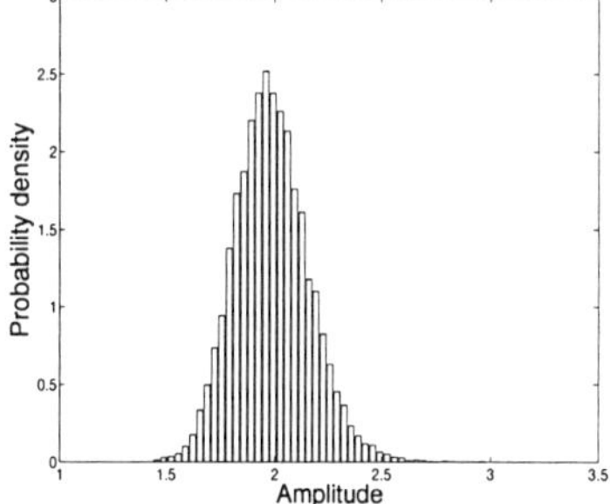

FIGURE 7.27. Two decay model: second amplitude

Table 7.5 shows the parameter estimates obtained using three different methods. The first column consists of estimates obtained by the experimentalist using the Marquardt algorithm, a variant of the curfit routine [12]. Note the absence of error bars. The second column consists of the most probable sample vector obtained during the course of the HMC run. The third column shows the results of a non-linear least squares analysis using a simplex algorithm with Gauss-Newton modification.

7.2.3 Concluding remarks

HMC has proven to be quite a remarkable technique. Its most notable feature is that it is a robust method of optimization. For the Lanczos prob-

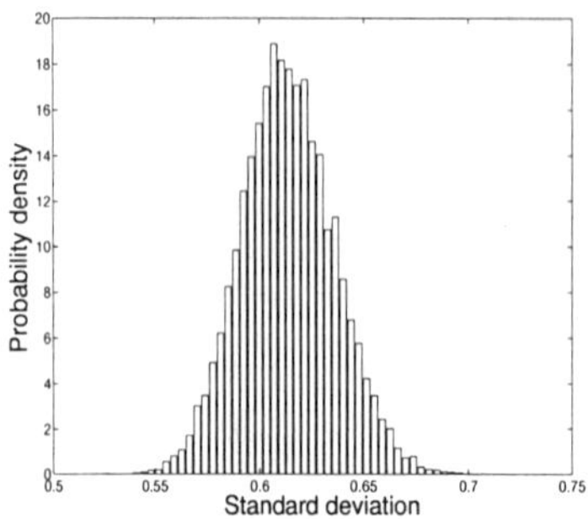

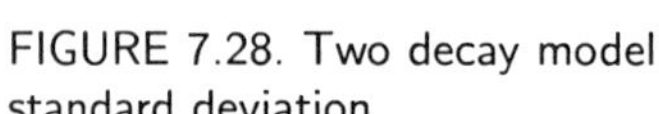
FIGURE 7.28. Two decay model: standard deviation

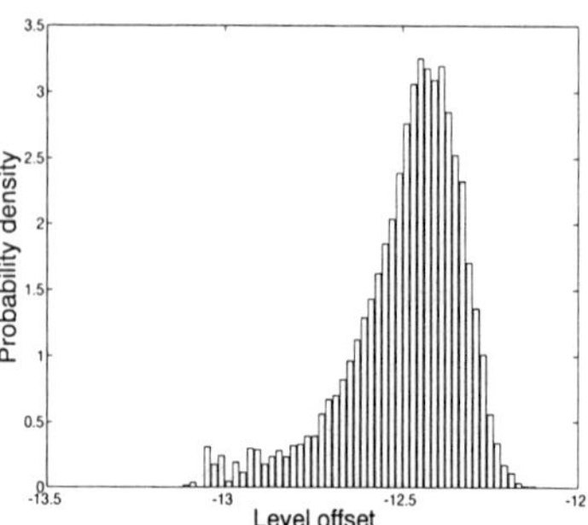

FIGURE 7.29. Two decay model: baseline

Biomedical Data	Parameter estimates		
	Marquardt	HMC	Non-linear LS
A_1	14.110	14.113	14.138 ± 0.47
A_2	1.9170	1.9296	1.9250 ± 0.15
λ_1	-4.5720	-4.5490	-4.5761 ± 0.27
λ_2	-0.2225	-0.2026	-0.2166 ± 0.05
c	-12.500	-12.516	-12.480 ± 0.12
σ	–	0.6081	–

TABLE 7.5. Parameter estimates for biomedical decay data

lem, a standard BFGS optimization routine did not converge and became numerically unstable. This occurred for most starting points. The HMC algorithm, on the other hand, never failed to converge. This was accomplished by inserting an additional check into the program, which meant that a point would be automatically rejected if the condition number of the matrix $\mathbf{G}^{\mathbf{T}}\mathbf{G}$ was dangerously high. Unlike deterministic optimization algorithms, the HMC trajectory joining any two points in parameter space is not unique so the HMC algorithm will effectively search for a stable trajectory. The final result produced by the HMC algorithm was in good agreement with the results obtained using the other techniques.

A second feature of the method is that once equilibrium is attained the algorithm produces a random sequence in which successive samples are almost independent. This makes the method particularly suitable for numerical integration.

There are some interesting points to note about the biomedical data. First, the data set was produced by data averaging, which means that the final data set is the mean of a number of so called "independent" realizations. Bretthorst [13] explains why data averaging is not a good idea. The most important reason is that, although the variance is reduced with data averaging, the resulting increase in resolution of the parameter estimates is a fraction of that which would be obtained if all the data were given separately. The Bayesian technique has no problem analyzing multiple realizations of the data at a single instant in time. Second, because of the nature of the experiment the reliability of the first few measurements is rather questionable. This view is borne out by a simple examination of the residual error in fitting the data.

The biomedical data set was not particularly difficult to analyse since the decays were well separated and there was a relatively large number of data points. It would be interesting to analyse other data sets where the decay rates are more closely spaced.

In this analysis, we have shown the possible benefits of a Bayesian analysis of the data. The analysis itself takes at most a few hours on a PC. The experimental equipment, on the other hand, is often expensive and the time taken in man-hours to gather the data can often be quite considerable. There are considerable benefits to be obtained using optimal Bayesian techniques over and above methods currently used in the experimental sciences.

7.3 General model selection

In this section, we present two examples of general model selection where we determine the most appropriate noise statistics and signal model from a set of possible models. The data is synthetic in each case.

Two signal models F2 and F2D, are proposed to fit the data.

$$\begin{aligned} \text{F2}: \quad f(t) &= A_1 \cos\omega_1 t + B_1 \sin\omega_1 t \\ &\quad + A_2 \cos\omega_2 t + B_2 \sin\omega_2 t \end{aligned} \tag{7.14}$$

$$\begin{aligned} \text{F2D}: \quad f(t) &= (A_1 \cos\omega_1 t + B_1 \sin\omega_1 t \\ &\quad + A_2 \cos\omega_2 t + B_2 \sin\omega_2 t) \exp(-\alpha t) \end{aligned} \tag{7.15}$$

The model F2 is clearly a special case of F2D.

Two possible noise models are also proposed for the noise statistics. These are the Gaussian and Laplacian noise assumptions respectively. This gives a total of four possible data models.

Uniform priors are assumed for the amplitude and decay location parameters and log-uniform (Jeffreys') priors are assumed for the frequency and noise variance scale parameters. The frequency and noise variance scale parameters were also transformed using a log transformation to map them onto the real axis.

The model evidences may be computed using the procedure detailed in section 3.8. The stages in the analysis are:

- *Non-linear optimization using BFGS Variable Metric algorithm.*

- *Numerical differentiation of the log density at the mode.*

- *Monte Carlo importance sampling.*

In the case of Laplacian noise statistics, it was impossible to differentiate at the mode of the posterior density to obtain the Hessian matrix. The simple solution was to apply the Hessian matrix obtained for Gaussian statistics.

7.3.1 Model selection in an impulsive noise environment

Consider the data plotted in figure 7.30 which shows two hundred synthetic data points generated using model 7.14 and corrupted by Laplacian noise.

Monte Carlo integration (using four thousand samples) gives the evidences for the four possible data models as shown in table 7.6. The probabilities for the four data models are shown in table 7.7. The result is quite conclusive. The data is best explained (99.26%) as the sum of two frequencies in Laplacian noise.

	Evidence	
Model	*Laplacian*	*Gaussian*
F2	$4.71 \pm 0.34 \times 10^{-210}$	$9.70 \pm 0.24 \times 10^{-216}$
F2D	$3.48 \pm 0.23 \times 10^{-212}$	$6.35 \pm 0.40 \times 10^{-217}$

TABLE 7.6. Evidence for the four data models

	Rel. prob.	
Model	*Laplacian*	*Gaussian*
F2	99.26%	0.000206%
F2D	0.733%	0.000013%

TABLE 7.7. Relative probabilities for the four data models

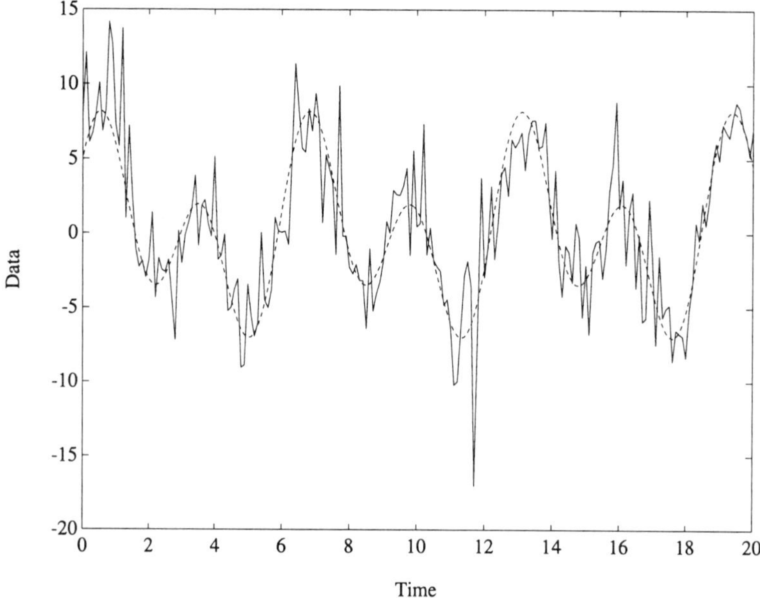

FIGURE 7.30. Two frequencies in Laplacian noise

7.3.2 Model selection in a Gaussian noise environment

Figure 7.31 shows two hundred synthetic data points generated using model 7.14 and corrupted by Gaussian noise.

Monte Carlo integration (using four thousand samples) gives the evidences for the four possible data models as shown in table 7.8. The probabilities for the four data models are shown in table 7.9. Again the result is quite conclusive. The data is best explained (91.84%) as the sum of two frequencies in Gaussian noise.

Model	Evidence	
	Laplacian	*Gaussian*
F2	$3.25 \pm 0.27 \times 10^{-222}$	$6.66 \pm 0.30 \times 10^{-217}$
F2D	$6.62 \pm 2.78 \times 10^{-223}$	$5.91 \pm 0.69 \times 10^{-218}$

TABLE 7.8. Evidence for the four data models

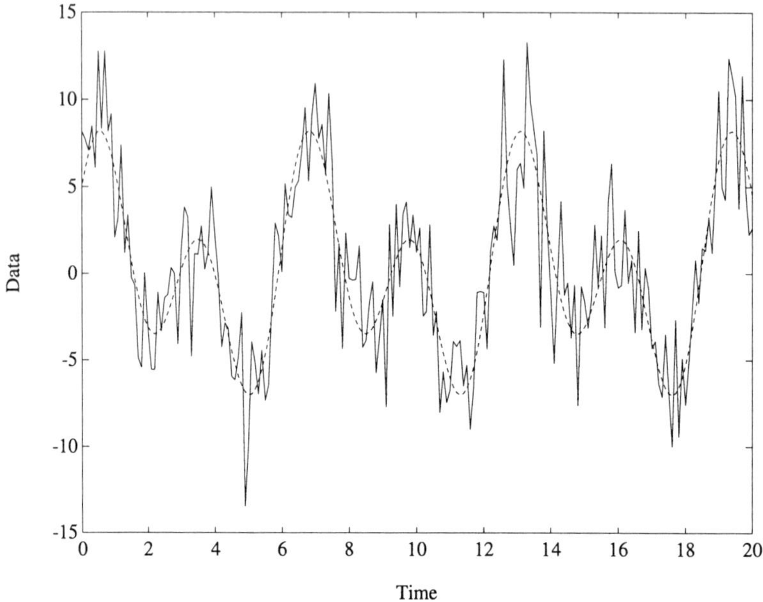

FIGURE 7.31. Two frequencies in Gaussian noise

	Rel. prob.	
Model	*Laplacian*	*Gaussian*
F2	0.000489%	91.84%
F2D	0.000091%	8.157%

TABLE 7.9. Relative probabilities for the four data models

7.4 Summary

This chapter has illustrated numerical approaches to the optimization and integration of the posterior density. In the first example polynomial data corrupted by Gaussian noise was analysed, by determining the model evidence and selected marginal densities. The analytical results were compared with numerical results obtained using a number of different techniques. In almost all cases the numerical methods performed well.

The power of the Hybrid Monte Carlo method was demonstrated by analyzing two data sets consisting of a superposition of decaying exponentials. The posterior mode was located without difficulty during the initial transitional period of the Markov chain. The marginal densities were obtained by forming histograms of components of the jointly distributed variates.

The model order was inferred for the Lanczos exponential data using a number of different numerical techniques to evaluate the model evidence. Two examples were also presented where model evidence was evaluated using importance sampling to determine the most appropriate choice of signal model and noise statistics. The correct model was recovered in all cases.

8 Conclusion

This book was concerned with devising numerical methods for practical applications of Bayesian methods to signal processing. There were three components in this work. First, we considered optimisation for the location of the posterior mode. Second, we investigated different strategies for integrating the posterior density. Third, we attempted to simulate random samples from the posterior density.

8.1 A review of the work

The book began with an introduction that presented simple models for which marginal densities and evidence could be obtained in close form. More general problems were then considered where statistics cannot be evaluated directly and can only be approximated using numerical techniques.

Chapter 2 began with a discussion of Bayesian methods, such as the application of marginalisation to parameter estimation and of Bayesian evidence to model selection. The general linear model was described in some detail. In the examples that we discussed, the posterior density could either be integrated directly in closed form or be well approximated by a closed form expression. Examples were given to show the method in action.

Unfortunately however, not all problems are linear in the parameters and involve Gaussian additive noise. It is clear that the so called "general linear model" is quite specialised. For instance, a simple example that can present difficulty is non-Gaussian noise corruption of data. The extension of the Bayesian formalism given in chapter 2 to more general classes of signal model requires specialised numerical techniques.

Chapter 3 presented a detailed exposition of the strategies for numerical Bayesian inference. We began with a discussion of the principle of asymptotic normality which gives a guide to the behaviour of the posterior density in the presence of many data. We also discussed the rôle that parameterisation plays in facilitating the use of numerical techniques. Optimisation using local and global algorithms was also discussed. The strategies mentioned here could be implemented for the purpose of obtaining MAP estimates. However, our main aim in discussing optimisation was as a prelude to numerical integration.

There are many strategies available for solving integration problems in Bayesian time series analysis. The three main classes of method discussed were numerical quadrature, asymptotic approximations and Monte Carlo integration. Sampling based methods in particular proved highly suitable for solving higher dimensional problems. The principle of importance sampling was motivated by the premise that we should, as far as possible, sample from the integrand when performing an integral. The usefulness of being able to simulate random samples from the joint density became obvious.

Based on these principles, an effective method for determining model evidence was devised. We also introduced methods for computing marginal densities, the simplest approach being to use histograms. We also developed the dummy variable method for computing marginal densities. The technique of averaging the conditional density using jointly distributed variates was shown to be a special case of the dummy variable method.

Chapter 4 was devoted to Markov chain Monte Carlo methods. There are many Markov chain based methods for sampling from a distribution. Two of the most common are the Metropolis algorithm and the Gibbs sampler. In addition there are the dynamical methods such as stochastic dynamics and the Hybrid Monte Carlo method. A fundamental step in the development of these methods is the establishment of a link between the concepts of an energy function and probability density. By introducing a hypothetical time coordinate and by tracing the dynamics of the system one has a remarkably powerful method of generating states drawn from the canonical distribution.

Simulated annealing was presented as a means of locating the dominant posterior mode (the global maximum) in parameter space in preference to spurious local maxima. A new approach to simulated annealing based on the Hybrid Monte Carlo method was discussed. We also discussed free energy methods, such as thermodynamic integration, which gave rise to novel methods for determining Bayesian evidence.

The approaches described above were devised for practical applications of the Bayesian formalism involving non-Gaussian noise models as well as non-linear parameters. Several diverse applications have been presented:

Chapter 5 developed an approach to the detection of abrupt discontinuities in time series. A model known as the general piecewise linear model was devised for determining sudden changes in the linear parameters of a model. Applications include the location of abrupt changes in the mean of a Gaussian process, as well as changes in the coefficients of an autoregressive model. The location of multiple changepoints in both Gaussian and non-Gaussian noise was investigated. The parameter space was too large and the likelihood function too complicated for an exhaustive search through parameter space. Therefore, an approach using simulated annealing and the Gibbs sampler was devised to overcome this problem and to determine a near-optimal solution.

Chapter 6 investigated a very interesting example of where random simulation of the posterior density could be put to real practical use. The problem considered was the restoration of missing samples for click removal in digital audio signals. An autoregressive model was assumed for the signal. Partitioning the parameter space into missing data, AR parameters and standard deviation led to a very elegant solution using the Gibbs sampler. It was found that simulating missing values in this way led to robust global optimisation, as well as preservation of the excitation in the gap. The comparison with the results obtained using the ML and EM algorithms, which were also developed in this chapter, were very favourable.

Chapter 7 applied several integration strategies to a number of different problems. Numerical optimisation was also often important in these applications. In the polynomial example, different methods for computing evidence and computing marginal densities were tested by comparing the numerical results with known analytical forms. We also tackled the very difficult (some would say infamous) problem of determining the number of decaying exponentials in decay data, as well as their decay rates. The Hybrid Monte Carlo algorithm was used to do this. Starting at some arbitrary initial position, the HMC algorithm converged rapidly to equilibrium. Thereafter the parameters could be estimated, jointly from the samples, or individually by forming histograms to estimate the marginal densities. The method was applied to biomedical data. The shapes of the marginal densities obtained were often skewed. Furthermore, the samples could be plotted against each other to form two dimensional marginal densities. The resulting probability cloud often had a very interesting crescent shape.

8.2 Further work

Yates [140] applied the GPL model to the detection of kickback in oil well data. Kickback is a build up of gas pressure at the well head which can have potentially disastrous consequences to life and property. The standard approach to such a problem was to apply the Hinckley test which, as we have seen, is the optimal detector for changes in the mean of a Gaussian process. Examination of the data shows that kickback is in fact characterized by an abrupt (locally) linear increase in pressure. Yates [140] derived an optimal Bayesian detector to detect ramps which, when applied to real data, gave a very significant improvement in the time in which kickback could be detected. This is an on-line application, rather than a retrospective analysis, so the techniques used in chapter 5 are not wholly appropriate. Further work on on-line detection of this phenomenon could proceed using a linear dynamic model approach similar to that proposed by Gordon [44].

Stephens and Smith [123] have applied the one dimensional simple step detector to edge detection in images with reasonable success. Their approach was to locate changepoints along the vertical and horizontal directions to cross validate the results. A proper formulation of the changepoint problem in two dimensions should produce an algorithm that is far more effective.

The Bayesian approach used in chapter 5 has the great disadvantage that the number of changepoints must be specified in advance of analysing the data. This problem could be addressed by carrying out a preliminary model order selection stage to determine the appropriate number of changepoints. McCulloch and Tsay [76] and Stephens [121, 122] provide an arguably superior approach in which the selection of the number of changepoints is carried out as part of the Gibbs sampling scheme.

In chapter 5, we also considered the location of changepoints in non-Gaussian Laplacian noise. Carlin, Gelfand and Smith [21] analyse non-Gaussian data corrupted by Poisson noise by means of the Gibbs sampler. Poisson processes commonly occur in experimental data and it is certain that many further applications can be found.

Curiously, as far as we are aware, there are no examples in the literature of the use of simulated annealing for this kind of problem. Future work with simulated annealing can proceed on many fronts. The annealing schedules used for simulated annealing can also be greatly improved. Options include adaptive annealing schedules such as using a constant rate of entropy reduction [88] or a constant thermodynamic speed [110] as described in section 4.7.1. In addition to these modifications, one could also employ Hybrid Monte Carlo method as described in section 4.6.5 to propose new states for continuous parameters in cases where the pdf is differentiable.

The Gibbs sampling approach devised in chapter 6 can be easily combined with the changepoint detection techniques of McCulloch and Tsay [76]. This presents the exciting possibility of combining click detection in audio signals with the restoration of the missing data left behind after the clicks are removed. Preliminary work carried out by Godsill and Rayner [41] has given very encouraging results.

Because Gibbs sampling maintains an excitation in the gap, it may be used to interpolate over gaps of any length without limit, providing the assumptions made about the signal model remain true (e.g. AR model, fixed model order, stationarity etc.). Rajan and Rayner [101] and Rajan, Rayner and Godsill [100] have also developed methods for the interpolation of *non-stationary* autoregressive processes using the Gibbs sampler. One example they present is the successful interpolation of a chirped sine wave with added Gaussian noise. This kind of signal cannot be processed using the methods in chapter 6. Future work may extend the application of the Gibbs sampler for parameter estimation and interpolation in data sets other than those described by autoregressive models.

The Hybrid Monte Carlo method is a rather exciting development. Previously the method was applied in physics for carrying out random simulations of physical systems. In this book, it was shown how Hybrid Monte Carlo may be used for general data analysis.

One possible application of the HMC method would be to carry out spectral analysis using simulated annealing. The energy function for a multiple frequency model is highly multimodal, so the location of the minimum is non-trivial. The gradient information is not directly as useful in this application as it was in the decaying exponential model. In the worst case the kinetic energy simply behaves as a source of energy which enables the trajectory to escape from local minima. This is still useful since it gives an annealing schedule that exhibits properties that are something similar, if not identical, to constant entropy reduction. Initial results using simulated data have been encouraging. Decaying sinusoidal components that characterize many linear second order systems could also be analysed in this way. It may even be possible to apply the HMC algorithm to problems involving discrete pdfs which are themselves quantized versions of continuous pdfs (e.g. the Poisson pdf may be viewed as a quantized gamma pdf). In this book, we have shown that there are potentially many new applications to be found in data analysis.

The principle of asymptotic normality was extensively applied throughout this work. In some circumstances it was applied directly. In general this required some non-linear optimisation technique to locate the posterior mode and numerical differentiation to obtain the covariance matrix. Looser applications based on the principle of asymptotic normality appeared, such as in devising an importance sampling strategy to obtain the evidence, as well as selecting auxiliary importance sampling densities to be used in the dummy variable method. Some applications were almost independent of the normal approximation. The Hybrid Monte Carlo method did not require the normal approximation, except that the appropriate choices for the leapfrog step size and the number of steps that are required when applying the algorithm to a specific problem were often deduced from a Gaussian approximation to the posterior density. The Gibbs sampling approach to digital audio restoration did not use asymptotic normality at all.

There are a number of reasons why one would like to remove the dependence on asymptotic normality. Foremost amongst these is that in really large scale problems one does not really want to have to work with the Hessian matrix because of the expense both in terms of computation and computer memory. In addition some method of reparameterisation that does not depend explicitly on calculating the Hessian matrix would be desirable. Powell's method, conjugate gradients or variable metric, may hold the key to carrying this out. The HMC algorithm could be applied to large dimensional problems with ease. Some means of finding the appropriate step size and the number of leapfrog steps adaptively would be very useful.

There is much current interest within both the Statistics and Applied Sciences Communities in using numerical Markov Chain Monte Carlo methods. Recently, there have been several new developments which are already yielding very promising results. As described in the text, in order to carry out model selection, one calculates the evidence for each model separately using a Monte Carlo estimate of the evidence. It would clearly be more conceptually elegant if the model determination could be included in the Monte Carlo simulation of the parameter values for each model. This involves the construction of an algorithm which samples from a space of varying dimensionality.

Grenander and Miller [46] introduce a jump diffusion sampler in which transitions between models of different orders are made at random times and between jumps the parameters are simulated to satisfy a stochastic differential equation. Phillips and Smith [92] have used this sampling method for object recognition, with very good results. The reversible jump Markov chain Monte Carlo framework introduced by Green [45] provides a method of moving between the different models in a very effective way. Morris and Fitzgerald [78] have used this approach to model line scratches in degraded motion pictures. Jump method techniques promise to have important consequences and applications.

Appendix A
The General Linear Model

We shall consider the broad class of models in the form given by equation 2.23. This class includes AR models as used for audio restoration in chapter 6, changepoints as shown in chapter 5 as well as Bayesian spectral analysis as described in chapter 2.

The Gaussian likelihood function is:

$$p(\mathbf{d} \mid \{\omega\}, \sigma, \mathbf{b}, \mathbf{I}) = \left(2\pi\sigma^2\right)^{-\frac{N}{2}} \exp\left[-\frac{\mathbf{e}^{\mathbf{T}}\mathbf{e}}{2\sigma^2}\right] \tag{A.1}$$

where $\{\omega\}$ denotes the parameters embedded in the matrix of basis functions $\mathbf{G}$.

The parameters may be classified into three different categories:

- Model amplitudes $\mathbf{b}$.
- Standard deviation σ.
- Key parameters $\{\omega\}$.

The aim in this appendix is to detail the steps needed to evaluate the marginal density of the key parameters $\{\omega\}$ in closed form[1]

$$p(\{\omega\} \mid \mathbf{d}, \mathbf{I}) = \int_0^\infty \int_{\Re^M} p(\{\omega\}, \sigma, \mathbf{b} \mid \mathbf{d}, \mathbf{I})\, d\mathbf{b}\, d\sigma \tag{A.2}$$

---

[1]See Fitzgerald [32] for another derivation.

Assigning uniform priors to the linear parameters $\mathbf{b}$ and Jeffreys' prior to σ, the joint posterior density can be expressed in terms of the likelihood as:

$$p(\{\omega\}, \sigma, \mathbf{b} \mid \mathbf{d}, \mathbf{I}) \propto p(\mathbf{d} \mid \{\omega\}, \sigma, \mathbf{b}, \mathbf{I}) \frac{1}{\sigma} \tag{A.3}$$

Substituting equation 2.23 into equation A.1 gives

$$p(\{\omega\}, \sigma, \mathbf{b} \mid \mathbf{d}, \mathbf{I}) \propto \left(2\pi\sigma^2\right)^{-\frac{N}{2}} \exp\left[-\frac{(\mathbf{d}-\mathbf{G}\mathbf{b})^{\mathbf{T}}(\mathbf{d}-\mathbf{G}\mathbf{b})}{2\sigma^2}\right]\frac{1}{\sigma} \tag{A.4}$$

The marginal density for the key parameters may be obtained in closed form by integrating in two stages:

- Integrate out the model amplitudes $\mathbf{b}$.
- Integrate out the standard deviation σ.

Obtaining a conditional density is a related problem that may be solved by first obtaining the marginal density in closed form, and second, by dividing the joint density by the corresponding marginal density.

A.1 Integrating out model amplitudes

We wish to compute

$$p(\{\omega\}, \sigma \mid \mathbf{d}, \mathbf{I}) = \int_{\Re^M} p(\{\omega\}, \sigma, \mathbf{b} \mid \mathbf{d}, \mathbf{I})\, d\mathbf{b} \tag{A.5}$$

From the point of view of integrating out the model amplitudes $\mathbf{b}$ equation A.4 is the exponential of a negative definite quadratic form.

The quadratic term of interest is

$$(\mathbf{d}-\mathbf{G}\mathbf{b})^{\mathbf{T}}(\mathbf{d}-\mathbf{G}\mathbf{b}) = \mathbf{d}^{\mathbf{T}}\mathbf{d} - 2\mathbf{d}^{\mathbf{T}}\mathbf{G}\mathbf{b} + \mathbf{b}^{\mathbf{T}}\mathbf{G}^{\mathbf{T}}\mathbf{G}\mathbf{b} \tag{A.6}$$

The presence of cross terms $b_j b_i$ in the integral make it difficult to integrate the posterior density directly. The cross terms may be removed by a change of variables as described below.

A.1.1 Least squares

Clearly maximizing expression A.4 with respect to $\mathbf{b}$ is equivalent to minimizing expresssion A.6 with respect to $\mathbf{b}$.

$$\frac{\partial}{\partial \mathbf{b}}\mathbf{e}^{\mathbf{T}}\mathbf{e} = 0 \tag{A.7}$$

Solving gives $\hat{\mathbf{b}} = \left(\mathbf{G}^{\mathbf{T}}\mathbf{G}\right)^{-1}\mathbf{G}^{\mathbf{T}}\mathbf{d}$. This yields the least squares fit to the data $\mathbf{d}$.

$$\mathbf{f} = \mathbf{G}\hat{\mathbf{b}} = \mathbf{G}\left(\mathbf{G}^{\mathbf{T}}\mathbf{G}\right)^{-1}\mathbf{G}^{\mathbf{T}}\mathbf{d} \tag{A.8}$$

A.1.2 Orthogonalization

In this section, we will apply an orthogonalizing transformation to diagonalise the matrix $\mathbf{G}^{\mathbf{T}}\mathbf{G}$, and hence remove cross terms of the form $b_i b_j$ from the integral.

Let us write

$$\mathbf{b} = \mathbf{P}\mathbf{a} \tag{A.9}$$

where $\mathbf{G}^{\mathbf{T}}\mathbf{G} = \mathbf{P}\mathbf{\Lambda}\mathbf{P}^{\mathbf{T}}$ is the eigendecomposition[2]. Also let $\hat{\mathbf{b}} = \mathbf{P}\hat{\mathbf{a}}$.

Let us compare

$$c + (\mathbf{a} - \hat{\mathbf{a}})^{\mathbf{T}}\mathbf{\Lambda}(\mathbf{a} - \hat{\mathbf{a}}) = (\mathbf{d} - \mathbf{G}\mathbf{b})^{\mathbf{T}}(\mathbf{d} - \mathbf{G}\mathbf{b}) \tag{A.10}$$

which gives

$$\begin{aligned} c &= \left[\mathbf{d}^{\mathbf{T}}\mathbf{d} - \mathbf{d}^{\mathbf{T}}\mathbf{G}\left(\mathbf{G}^{\mathbf{T}}\mathbf{G}\right)^{-1}\mathbf{G}^{\mathbf{T}}\mathbf{d}\right] \\ &= \left[\mathbf{d}^{\mathbf{T}}\mathbf{d} - \mathbf{f}^{\mathbf{T}}\mathbf{f}\right] \end{aligned}$$

Integral A.5 becomes

$$\begin{aligned} p(\{\omega\}, \sigma \mid \mathbf{d}, \mathbf{I}) \propto \int_{\Re^M} &(2\pi\sigma^2)^{-\frac{N}{2}} \exp\left[-\frac{c}{2\sigma^2}\right] \times \\ &\exp\left[-\frac{(\mathbf{a} - \hat{\mathbf{a}})^{\mathbf{T}}\mathbf{\Lambda}(\mathbf{a} - \hat{\mathbf{a}})}{2\sigma^2}\right] d\mathbf{a} \end{aligned} \tag{A.11}$$

Note that $d\mathbf{b} = |\mathbf{P}|\, d\mathbf{a}$ and that $|\mathbf{P}| = 1$ because $\mathbf{P}$ is an orthonormal matrix.

The next stage in computing the integral is to note that

$$(\mathbf{a} - \hat{\mathbf{a}})^{\mathbf{T}}\mathbf{\Lambda}(\mathbf{a} - \hat{\mathbf{a}}) = \sum_{i=1}^{M} \lambda_i (a_i - \hat{a}_i)^2 \tag{A.12}$$

where λ_i are the eigenvalues.

Moreover

$$\begin{aligned} \int_{\Re^M} \exp\left[-\frac{\sum_{i=1}^{M} \lambda_i (a_i - \hat{a}_i)^2}{2\sigma^2}\right] d\mathbf{a} &= \prod_{i=1}^{M} \int_{-\infty}^{\infty} \exp\left[-\frac{\lambda_i (a_i - \hat{a}_i)^2}{2\sigma^2}\right] da_i \\ &= \prod_{i=1}^{M} \sqrt{\frac{2\pi\sigma^2}{\lambda_i}} = \frac{(2\pi\sigma^2)^{M/2}}{\sqrt{\prod_{i=1}^{M} \lambda_i}} \end{aligned}$$

---

[2]Note that $\mathbf{P}^{\mathbf{T}} = \mathbf{P}^{-1}$ because $\mathbf{G}^{\mathbf{T}}\mathbf{G}$ is a symmetric matrix.

The last stage is to see that $\prod_{i=1}^{M} \lambda_i = \left|\mathbf{G}^{\mathbf{T}}\mathbf{G}\right|$. This follows from the fact that $\left|\mathbf{G}^{\mathbf{T}}\mathbf{G}\right| = \left|\mathbf{P}\mathbf{\Lambda}\mathbf{P}^{\mathbf{T}}\right| = |\mathbf{P}||\mathbf{\Lambda}|\left|\mathbf{P}^{\mathbf{T}}\right| = |\mathbf{\Lambda}| = \prod_{i=1}^{M} \lambda_i$.

Finally we obtain

$$p(\{\omega\}, \sigma \mid \mathbf{d}, \mathbf{I}) \propto \frac{\left(2\pi\sigma^2\right)^{-(N-M)/2}}{\left|\mathbf{G}^{\mathbf{T}}\mathbf{G}\right|^{1/2}} \exp\left[-\frac{\left(\mathbf{d}^{\mathbf{T}}\mathbf{d} - \mathbf{f}^{\mathbf{T}}\mathbf{f}\right)}{2\sigma^2}\right] \frac{1}{\sigma} \tag{A.13}$$

A.2 Integrating out the standard deviation

Integrating out the remaining nuisance parameter, the standard deviation is relatively straightforward. It can be integrated out quite readily using the gamma integral in equation 2.27 to obtain:

$$p(\{\omega\} \mid \mathbf{d}, \mathbf{I}) \propto \frac{\frac{1}{2}\Gamma\left(\frac{N-M}{2}\right)}{\sqrt{\det\left(\mathbf{G}^{\mathbf{T}}\mathbf{G}\right)}\left[\pi\left(\mathbf{d}^{\mathbf{T}}\mathbf{d} - \mathbf{f}^{\mathbf{T}}\mathbf{f}\right)\right]^{\frac{N-M}{2}}} \tag{A.14}$$

For the remainder of this chapter we shall write the term in $\mathbf{f}^{\mathbf{T}}\mathbf{f}$ using a more commonly encountered equivalent:

$$\mathbf{f}^{\mathbf{T}}\mathbf{f} = \mathbf{d}^{\mathbf{T}}\mathbf{G}\left(\mathbf{G}^{\mathbf{T}}\mathbf{G}\right)^{-1}\mathbf{G}^{\mathbf{T}}\mathbf{d} \tag{A.15}$$

A.3 Marginal density for a linear coefficient

Suppose we require the marginal density of b_j (i.e. the jth of vector $\mathbf{b}$). We can rewrite equation 2.23 in the form:

$$\mathbf{d} - b_j\,\mathbf{c} = \mathbf{K}\,\mathbf{a} + \mathbf{e} \tag{A.16}$$

where $\mathbf{c}$ is the jth column of $\mathbf{G}$ such that $c_i = G_{ij}$, $\mathbf{K}$ is matrix $\mathbf{G}$ with the jth column removed and $\mathbf{a}$ is vector $\mathbf{b}$ with element b_j removed.

We will treat the vector $\mathbf{a}$ on the right hand side of equation A.16 as a vector of nuisance parameters. The left hand side of equation A.16 is a simple modification to the data vector:

$$\mathbf{r} = \mathbf{d} - b_j\,\mathbf{c} \tag{A.17}$$

We may integrate out the nuisance parameters $\mathbf{a}$ and the standard deviation σ using the methods given above. The marginal density will take the form of a student-t distribution as in equation A.14 above:

$$p(b_j \mid \mathbf{d}, \mathbf{I}) \propto \frac{\left[\mathbf{r}^{\mathbf{T}}\mathbf{r} - \mathbf{r}^{\mathbf{T}}\mathbf{K}\left(\mathbf{K}^{\mathbf{T}}\mathbf{K}\right)^{-1}\mathbf{K}^{\mathbf{T}}\mathbf{r}\right]^{\frac{-(N-M+1)}{2}}}{\sqrt{\det\left(\mathbf{K}^{\mathbf{T}}\mathbf{K}\right)}} \tag{A.18}$$

Note that the power is $(N - M + 1)/2$ rather than $(N - M)/2$, as one might expect from examining equation A.14, because there are only $M - 1$ elements in the vector $\mathbf{a}$.

A.4 Marginal density for standard deviation

The posterior density is given by:

$$p(\sigma, \mathbf{b} \mid \mathbf{d}, \mathbf{I}) \propto \sigma^{-N-1} \exp\left[-\frac{1}{2\sigma^2}(\mathbf{d} - \mathbf{Gb})^{\mathbf{T}}(\mathbf{d} - \mathbf{Gb})\right] \tag{A.19}$$

The parameters $\mathbf{b}$ may be integrated out as shown in section A.1 to give:

$$p(\sigma \mid \mathbf{d}, \mathbf{I}) \propto \sigma^{-N+M-1} \exp\left[-\frac{1}{2\sigma^2}\left(\mathbf{d}^{\mathbf{T}}\mathbf{d} - \mathbf{d}^{\mathbf{T}}\mathbf{G}\left(\mathbf{G}^{\mathbf{T}}\mathbf{G}\right)^{-1}\mathbf{G}^{\mathbf{T}}\mathbf{d}\right)\right] \tag{A.20}$$

This is in the form of an inverse chi density.

A.5 Conditional density for a linear coefficient

Let us partition the amplitude vector $\mathbf{b}$ into two components $\mathbf{x}$ and $\mathbf{y}$. We require the conditional density $p(\mathbf{x} \mid \mathbf{y}, \sigma, \mathbf{d}, \mathbf{I})$.

$$p(\mathbf{x} \mid \mathbf{y}, \sigma, \mathbf{d}, \mathbf{I}) = \frac{p(\mathbf{x}, \mathbf{y} \mid \sigma, \mathbf{d}, \mathbf{I})}{p(\mathbf{y} \mid \sigma, \mathbf{d}, \mathbf{I})} \tag{A.21}$$

Let us expand the signal in terms of its components:

$$\mathbf{Gb} = \mathbf{Kx} + \mathbf{Ly} \tag{A.22}$$

$\mathbf{K}$ and $\mathbf{L}$ are partitions of $\mathbf{G}$ consisting of the columns corresponding to $\mathbf{x}$ and $\mathbf{y}$ respectively.

It is easy to show (using the same approach as in section 6.4.2) that the conditional density is of the following form:

$$p(\mathbf{x} \mid \mathbf{y}, \sigma, \mathbf{d}, \mathbf{I}) \propto \sigma^{-p} \exp\left[-\frac{1}{2\sigma^2}(\mathbf{x} - \hat{\mathbf{x}})^{\mathbf{T}}\mathbf{K}^{\mathbf{T}}\mathbf{K}(\mathbf{x} - \hat{\mathbf{x}})\right] \tag{A.23}$$

where

$$\hat{\mathbf{x}} = \hat{\mathbf{x}}(\mathbf{y}) = \left(\mathbf{K}^{\mathbf{T}}\mathbf{K}\right)^{-\mathbf{1}}\mathbf{K}^{\mathbf{T}}(\mathbf{d} - \mathbf{L}\mathbf{y}) \tag{A.24}$$

and where the vector $\mathbf{y}$ is of length p. The conditional density is in a very convenient multivariate Gaussian form. The mean of the density is given by equation A.24 and the covariance matrix is:

$$\mathbf{C}_{\mathbf{X}} = \sigma^2\left(\mathbf{K}^{\mathbf{T}}\mathbf{K}\right)^{-1} \tag{A.25}$$

A.6 Conditional density for standard deviation

The joint density can be written in the form:

$$p(\sigma, \mathbf{b} \mid \mathbf{d}, \mathbf{I}) \propto \sigma^{-N-1} \exp\left[-\frac{Q}{\sigma^2}\right] \tag{A.26}$$

where

$$Q = \frac{1}{2}(\mathbf{d} - \mathbf{G}\mathbf{b})^{\mathbf{T}}(\mathbf{d} - \mathbf{G}\mathbf{b}) \tag{A.27}$$

Using the gamma integral in equation 2.27 we can show that:

$$p(\mathbf{b} \mid \mathbf{d}, \mathbf{I}) \propto Q^{-N/2} \tag{A.28}$$

Therefore the required conditional density is given by:

$$p(\sigma \mid \mathbf{b}, \mathbf{d}, \mathbf{I}) \propto Q^{N/2}\,\sigma^{-N-1} \exp\left[-\frac{Q}{\sigma^2}\right] \tag{A.29}$$

This is in the form of an inverse chi density.

Appendix B
Sampling from a Multivariate Gaussian Density

Suppose we are interested in generating random variates from a multivariate Gaussian density of the form

$$f_0(\mathbf{x}) = \frac{(2\,\pi)^{-m/2}}{\sqrt{\det \mathbf{C}}} \exp\left[-\frac{(\mathbf{x}-\mu)^{\mathbf{T}}\,\mathbf{C}^{-1}\,(\mathbf{x}-\mu)}{2}\right] \tag{B.1}$$

Let $\mathbf{n} = \mathbf{x} - \mu$ which gives

$$f_0(\mathbf{n}) = \frac{(2\,\pi)^{-m/2}}{\sqrt{\det \mathbf{C}}} \exp\left[-\frac{\mathbf{n}^{\mathbf{T}}\,\mathbf{C}^{-1}\,\mathbf{n}}{2}\right] \tag{B.2}$$

We are seeking some means of transforming a zero mean unit variance white Gaussian random vector into a random vector with the density given in equation B.2.

Consider the density of a unit variance Gaussian random vector

$$f_1(\mathbf{e}) = (2\,\pi)^{-m/2} \exp\left[-\frac{\mathbf{e}^{\mathbf{T}}\,\mathbf{e}}{2}\right] \tag{B.3}$$

Comparing equation B.2 with equation B.3 (and ignoring scaling) gives the relation

$$\mathbf{e}^{\mathbf{T}}\,\mathbf{e} = \mathbf{n}^{\mathbf{T}}\,\mathbf{C}^{-1}\,\mathbf{n} \tag{B.4}$$

Let $\mathbf{C}^{-1} = \mathbf{S}^{\mathbf{T}}\,\mathbf{S}$ which implies

$$\mathbf{e} = \mathbf{S}\,\mathbf{n} \tag{B.5}$$

The required random variates are therefore generated using

$$\mathbf{x} = \mathbf{S}^{-1}\,\mathbf{e} + \mu \tag{B.6}$$

Note that the transformation requires:

- The generation of a unit variance Gaussian random vector.
- The computation of the square root of the covariance matrix.
- The computation of $\mathbf{S}^{-1}\,\mathbf{e}$.

The square root of a matrix is not unique. In addition, a matrix must be positive definite before a real valued square root can be found.

Different kinds of matrix square root seem to be appropriate for different problems. Two different kinds of square root are commonly used.

Eigendecomposition Let $\mathbf{C}^{-1} = \mathbf{P}\,\mathbf{\Lambda}\,\mathbf{P}^{\mathbf{T}}$ where $\mathbf{P}$ is the matrix whose columns consist of orthonormal eigenvectors, and where $\mathbf{\Lambda}$ is the diagonal matrix of the corresponding eigenvalues. Writing $\mathbf{C}^{-1} = \mathbf{S}^{\mathbf{T}}\,\mathbf{S}$ implies $\mathbf{S}^{\mathbf{T}} = \mathbf{P}\,\mathbf{\Lambda}^{1/2}$ is a solution.

The eigendecomposition square root is suitable for Monte Carlo integration. This is because the covariance matrix is symmetric which implies that eigenvectors form an orthonormal set of basis vectors. If one decides to sample the orthonormal vectors using (orthonormal) independent random variates, then this will be equivalent to using an eigendecomposition square root to colour the noise in equation B.6. The square roots of the corresponding eigenvalues provide the scaling of the random variates.

Cholesky The Cholesky decomposition is a very commonly encountered matrix square root. In its standard form, an upper triangular $\mathbf{S}$ is determined such that $\mathbf{C}^{-1} = \mathbf{S}^{\mathbf{T}}\,\mathbf{S}$.

We apply the Cholesky decomposition to audio restoration because the Cholesky decomposition of a band diagonal matrix is itself band diagonal with the same upper bandwidth. This gives the following advantages of over kinds of square root:

- Memory requirements are reduced.
- Less computation is required to work with the band diagonal Cholesky decomposition than with other matrix square roots.

In addition, the Cholesky decomposition (which is band diagonal) may also be appropriate to the problem of interpolating AR processes because of the localized dependence of each sample output with its neighbours.

Appendix C
Hybrid Monte Carlo Derivations

In this appendix, we will present the results that are required in the main text to implement the Hybrid Monte Carlo method. The prior probabilities are taken to be flat and uninformative, which implies that the posterior density is proportional to the likelihood.

C.1 Full Gaussian likelihood

In this section we derive the gradient of the energy function (i.e. the "force") for all the parameters of a full Gaussian likelihood.

The posterior density is given by the expression:

$$p(\{\omega\}, \mathbf{b}, \sigma \mid \mathbf{d}) \propto \left(2\pi\sigma^2\right)^{-N/2} \exp\left[-\frac{(\mathbf{d}-\mathbf{G}\mathbf{b})^{\mathbf{T}}(\mathbf{d}-\mathbf{G}\mathbf{b})}{2\sigma^2}\right] \tag{C.1}$$

Ignoring additive constants the energy function may be written in the form

$$\begin{aligned} E(\{\omega\}, \mathbf{b}, \sigma \mid \mathbf{d}) &= -\log p(\{\omega\} \mid \mathbf{d}) \\ &= N\log\sigma + \frac{(\mathbf{d}-\mathbf{G}\mathbf{b})^{\mathbf{T}}(\mathbf{d}-\mathbf{G}\mathbf{b})}{2\sigma^2} \end{aligned} \tag{C.2}$$

We will now differentiate the energy function with respect to $\mathbf{b}$, σ, and $\{\omega\}$ in turn.

1.
$$\frac{\partial E}{\partial \mathbf{b}} = -\frac{1}{\sigma^2}\mathbf{G}^{\mathbf{T}}(\mathbf{d}-\mathbf{G}\mathbf{b}) \tag{C.3}$$

2.
$$\frac{\partial E}{\partial \sigma} = \frac{N\sigma^2 - (\mathbf{d}-\mathbf{G}\mathbf{b})^{\mathbf{T}}(\mathbf{d}-\mathbf{G}\mathbf{b})}{\sigma^3} \tag{C.4}$$

3.
$$\frac{\partial E}{\partial \alpha} = -\frac{1}{\sigma^2}(\mathbf{d}-\mathbf{G}\mathbf{b})^{\mathbf{T}}\mathbf{L}\mathbf{b} \tag{C.5}$$

where α is a scalar component of $\{\omega\}$ and where $\mathbf{L} = \frac{\partial}{\partial\alpha}\mathbf{G}$. It is not possible to differentiate with respect to $\{\omega\}$ as a whole as for the linear parameters $\mathbf{b}$ because of the non-linear nature of the $\{\omega\}$ parameters.

C.2 Student-t distribution

In this section, we will apply vector calculus to derive an expression for the gradient of an energy function for the general linear model where the amplitude vector $\mathbf{b}$ and standard deviation have been integrated out.

The posterior density for the parameters of the model function matrix $\mathbf{G}$ is given expression A.14 as:

$$p(\{\omega\} \mid \mathbf{d}, \mathbf{I}) \propto \frac{\frac{1}{2}\Gamma\left(\frac{N-M}{2}\right)}{\sqrt{\det\left(\mathbf{G}^{\mathbf{T}}\mathbf{G}\right)}\left[\pi\left(\mathbf{d}^{\mathbf{T}}\mathbf{d}-\mathbf{f}^{\mathbf{T}}\mathbf{f}\right)\right]^{\frac{N-M}{2}}} \tag{C.6}$$

Ignoring additive constants the energy function may be written in the form

$$\begin{aligned} E(\{\omega\} \mid \mathbf{d}) &= -\log p(\{\omega\} \mid \mathbf{d}) \\ &= \frac{N-M}{2}\log\left[\mathbf{d}^{\mathbf{T}}\mathbf{d}-\mathbf{f}^{\mathbf{T}}\mathbf{f}\right] + \frac{1}{2}\log\det\mathbf{G}^{\mathbf{T}}\mathbf{G} \end{aligned} \tag{C.7}$$

where

$$\mathbf{f} = \mathbf{G}\left(\mathbf{G}^{\mathbf{T}}\mathbf{G}\right)^{-1}\mathbf{G}^{\mathbf{T}}\mathbf{d}$$

Let $\mathbf{b}$ be defined as

$$\mathbf{b} = \left(\mathbf{G}^{\mathbf{T}}\mathbf{G}\right)^{-1}\mathbf{G}^{\mathbf{T}}\mathbf{d}$$

Our aim is to calculate $\frac{\partial E}{\partial\alpha}$ where α is a scalar component of ω.

Two fundamental identities from vector calculus are required to compute this derivative:

1.
$$\frac{\partial}{\partial\alpha}\mathbf{A}^{-1} = -\mathbf{A}^{-1}\left(\frac{\partial}{\partial\alpha}\mathbf{A}\right)\mathbf{A}^{-1} \tag{C.8}$$

2.
$$\frac{\partial}{\partial\alpha}\log\det\mathbf{A} = \text{Trace}\left(\mathbf{A}^{-1}\frac{\partial}{\partial\alpha}\mathbf{A}\right) \tag{C.9}$$

Differentiating expression C.7 above yields

$$\begin{aligned}\frac{\partial E}{\partial\alpha} &= \frac{N-M}{2}\left[\mathbf{d}^{\mathbf{T}}\mathbf{d}-\mathbf{f}^{\mathbf{T}}\mathbf{f}\right]^{-1}\left[-2\,\mathbf{f}^{\mathbf{T}}\frac{\partial\mathbf{f}}{\partial\alpha}\right]\\ &\quad+\frac{1}{2}\,\text{Trace}\left\{\left(\mathbf{G}^{\mathbf{T}}\mathbf{G}\right)^{-1}\left[\mathbf{G}^{\mathbf{T}}\mathbf{L}+\mathbf{L}^{\mathbf{T}}\mathbf{G}\right]\right\}\end{aligned} \tag{C.10}$$

where

$$\begin{aligned}\mathbf{L} &= \frac{\partial\mathbf{G}}{\partial\alpha}\\ \frac{\partial\mathbf{f}}{\partial\alpha} &= \mathbf{L}\left(\mathbf{G}^{\mathbf{T}}\mathbf{G}\right)^{-1}\mathbf{G}^{\mathbf{T}}\mathbf{d}+\mathbf{G}\left(\mathbf{G}^{\mathbf{T}}\mathbf{G}\right)^{-1}\mathbf{L}^{\mathbf{T}}\mathbf{d}\\ &\quad-\mathbf{G}\left(\mathbf{G}^{\mathbf{T}}\mathbf{G}\right)^{-1}\left[\mathbf{G}^{\mathbf{T}}\mathbf{L}+\mathbf{L}^{\mathbf{T}}\mathbf{G}\right]\left(\mathbf{G}^{\mathbf{T}}\mathbf{G}\right)^{-1}\mathbf{G}^{\mathbf{T}}\mathbf{d}\end{aligned}$$

After considerable algebraic manipulation one can show that:

$$\mathbf{f}^{\mathbf{T}}\frac{\partial\mathbf{f}}{\partial\alpha} = \mathbf{b}^{\mathbf{T}}\mathbf{L}^{\mathbf{T}}(\mathbf{d}-\mathbf{f}) = (\mathbf{d}-\mathbf{f})^{\mathbf{T}}\mathbf{L}\mathbf{b} \tag{C.11}$$

Finally one obtains

$$\frac{\partial E}{\partial\alpha} = -\frac{1}{\sigma^2}(\mathbf{d}-\mathbf{G}\mathbf{b})^{\mathbf{T}}\mathbf{L}\mathbf{b}+\text{Trace}\left\{\left(\mathbf{G}^{\mathbf{T}}\mathbf{G}\right)^{-1}\left[\mathbf{G}^{\mathbf{T}}\mathbf{L}\right]\right\} \tag{C.12}$$

where

$$\sigma^2 = \frac{1}{N-M}\left[\mathbf{d}^{\mathbf{T}}\mathbf{d}-\mathbf{f}^{\mathbf{T}}\mathbf{f}\right]$$

The similarity between equation C.5 and equation C.12 is very interesting.

C.3 Remark

The matrix $\mathbf{L} = \frac{\partial}{\partial\alpha}\mathbf{G}$ in equation C.5 and equation C.12 is more often than not very sparse; typically only a small number of columns of $\mathbf{G}$ depend on the parameter α and therefore only the corresponding columns of $\mathbf{L}$ will be non-zero. This may be easily exploited to give more efficient computation of the gradient of the energy function.

Appendix D
EM Algorithm Derivations

In this appendix, we present full derivations for terms used in the EM algorithm for audio restoration in chapter 6. The first derivation is for the Q function used in the expectation stage. The second derivation deals with an apparently awkward term involving the trace of a matrix (which is completely ignored by other authors [137]).

D.1 Expectation

The major difficulty in the expectation stage of the EM algorithm applied to autoregressive models is in evaluating the integral that defines the Q function in closed form.

For compactness, we will denote $\mathbf{L}_{i+1}$ as $\mathbf{L}$ and $\mathbf{L}_i$ as $\mathbf{M}$ in order to clarify the algebra in this section.

It is convenient to make the following substitutions:

$$\begin{aligned}(\mathbf{w}-\mathbf{L}\,\theta)^{\mathbf{T}}(\mathbf{w}-\mathbf{L}\,\theta) &= \left(\theta-\hat{\theta}\right)^{\mathbf{T}}\mathbf{L}^{\mathbf{T}}\mathbf{L}\left(\theta-\hat{\theta}\right)\\ &\quad+\mathbf{w}^{\mathbf{T}}\mathbf{w}-\mathbf{w}^{\mathbf{T}}\mathbf{L}\left(\mathbf{L}^{\mathbf{T}}\mathbf{L}\right)^{-1}\mathbf{L}^{\mathbf{T}}\mathbf{w}\end{aligned}$$

$$\begin{aligned}(\mathbf{v}-\mathbf{M}\,\theta)^{\mathbf{T}}(\mathbf{v}-\mathbf{M}\,\theta) &= \left(\theta-\hat{\phi}\right)^{\mathbf{T}}\mathbf{M}^{\mathbf{T}}\mathbf{M}\left(\theta-\hat{\phi}\right)\\ &\quad+\mathbf{v}^{\mathbf{T}}\mathbf{v}-\mathbf{v}^{\mathbf{T}}\mathbf{M}\left(\mathbf{M}^{\mathbf{T}}\mathbf{M}\right)^{-1}\mathbf{M}^{\mathbf{T}}\mathbf{v}\end{aligned}$$

where

$$\hat{\theta} = \left(\mathbf{L}^{\mathbf{T}}\,\mathbf{L}\right)^{-1}\mathbf{L}^{\mathbf{T}}\,\mathbf{w} \tag{D.1}$$

$$\hat{\phi} = \left(\mathbf{M}^{\mathbf{T}}\,\mathbf{M}\right)^{-1}\mathbf{M}^{\mathbf{T}}\,\mathbf{v} \tag{D.2}$$

Furthermore, let

$$\mathbf{p} = \theta - \hat{\phi} \tag{D.3}$$

and also let

$$\mathbf{q} = \mathbf{p} - \left(\hat{\theta} - \hat{\phi}\right) = \theta - \hat{\theta} \tag{D.4}$$

The Q function is therefore given by

$$Q(\mathbf{z}_{i+1}, \mathbf{z_i}) = \int_{\Theta} A\,(\mathbf{p})\,\exp\left[-B\,(\mathbf{p})\right] d\,\mathbf{p} \tag{D.5}$$

where

$$\begin{aligned} A\,(\mathbf{p}) \quad &= \quad -\frac{1}{2\,\sigma^2}\left[\mathbf{q}^{\mathbf{T}}\,\mathbf{L}^{\mathbf{T}}\mathbf{L}\mathbf{q} + \mathbf{w}^{\mathbf{T}}\,\mathbf{w} - \mathbf{w}^{\mathbf{T}}\,\mathbf{L}\left(\mathbf{L}^{\mathbf{T}}\,\mathbf{L}\right)^{-1}\mathbf{L}^{\mathbf{T}}\,\mathbf{w}\right] \\ &\quad -\frac{N}{2}\,\log\left(2\,\pi\,\sigma^2\right) \end{aligned}$$

$$\begin{aligned} B\,(\mathbf{p}) \quad &= \quad -\frac{1}{2\,\sigma^2}\left[\mathbf{p}^{\mathbf{T}}\,\mathbf{M}^{\mathbf{T}}\mathbf{M}\,\mathbf{p} + \mathbf{v}^{\mathbf{T}}\,\mathbf{v} - \mathbf{v}^{\mathbf{T}}\,\mathbf{M}\left(\mathbf{M}^{\mathbf{T}}\,\mathbf{M}\right)^{-1}\mathbf{M}^{\mathbf{T}}\,\mathbf{v}\right] \\ &\quad -\frac{N}{2}\,\log\left(2\,\pi\,\sigma^2\right) \end{aligned}$$

With a little algebraic manipulation one can simplify $A\,(\mathbf{p})$ to give

$$\begin{aligned} A\,(\mathbf{p}) \quad &= \quad -\frac{1}{2\,\sigma^2}\Big[\mathbf{p}^{\mathbf{T}}\,\mathbf{L}^{\mathbf{T}}\mathbf{L}\,\mathbf{p} - 2\,\mathbf{p}^{\mathbf{T}}\,\mathbf{L}^{\mathbf{T}}\mathbf{L}\left(\hat{\theta} - \hat{\phi}\right) \\ &\qquad + \left(\mathbf{w} - \mathbf{L}\,\hat{\phi}\right)^{\mathbf{T}}\left(\mathbf{w} - \mathbf{L}\,\hat{\phi}\right)\Big] - \frac{N}{2}\,\log\left(2\,\pi\,\sigma^2\right) \end{aligned}$$

The following integral identities are of use in evaluating the Q function:

$$\int_{\Re^q} \exp\left[-\mathbf{x}^{\mathbf{T}}\,\mathbf{B}\,\mathbf{x}\right] d\,\mathbf{x} = \frac{\pi^{\frac{q}{2}}}{\sqrt{\det \mathbf{B}}} \tag{D.6}$$

$$\int_{\Re^q} \mathbf{x}^{\mathbf{T}}\,\mathbf{A}\,\mathbf{y}\,\exp\left[-\mathbf{x}^{\mathbf{T}}\,\mathbf{B}\,\mathbf{x}\right] d\,\mathbf{x} = 0 \tag{D.7}$$

$$\int_{\Re^q} \mathbf{x}^{\mathbf{T}}\,\mathbf{A}\,\mathbf{x}\,\exp\left[-\mathbf{x}^{\mathbf{T}}\,\mathbf{B}\,\mathbf{x}\right] d\,\mathbf{x} = \frac{\pi^{\frac{q}{2}}}{\sqrt{\det \mathbf{B}}}\,\frac{\mathrm{Trace}\,(\mathbf{B}^{-1}\,\mathbf{A})}{2} \tag{D.8}$$

Note that identity D.7 assumes that $\mathbf{y}$ is independent of $\mathbf{x}$.

Using these integral identities, it is a straightforward calculation to obtain the Q function.

$$Q(\mathbf{z}_{i+1}, \mathbf{z_i}) = K(\mathbf{z}_i) \left[\frac{\text{Trace}\left[\left(\mathbf{M}^{\mathbf{T}}\mathbf{M}\right)^{-1} \left(\mathbf{L}^{\mathbf{T}}\mathbf{L}\right) \right]}{2} \right.$$
$$\left. + \frac{\left(\mathbf{w} - \mathbf{L}\hat{\phi}\right)^{\mathbf{T}} \left(\mathbf{w} - \mathbf{L}\hat{\phi}\right)}{2\sigma^2} - \frac{N}{2} \log\left(2\pi\sigma^2\right) \right]$$

where

$$K(\mathbf{z}_i) = -\frac{\left(2\pi\sigma^2\right)^{-\frac{N}{2}}}{\sqrt{\det \mathbf{M}^{\mathbf{T}}\mathbf{M}}} \exp\left[-\frac{\left(\mathbf{v}^{\mathbf{T}}\mathbf{v} - \mathbf{v}^{\mathbf{T}}\mathbf{M}\left(\mathbf{M}^{\mathbf{T}}\mathbf{M}\right)^{-1}\mathbf{M}^{\mathbf{T}}\mathbf{v}\right)}{2\sigma^2} \right] \tag{D.9}$$

D.2 Maximization

One major difficulty in the maximization stage of the EM algorithm is in differentiating the trace term in the Q function. We wish to determine $\mathbf{z}_{i+1}$ such that $Q(\mathbf{z}_{i+1}, \mathbf{z_i})$ is maximized. This gives:

$$\frac{\partial}{\partial \mathbf{z}_{i+1}} Q(\mathbf{z}_{i+1}, \mathbf{z_i}) = 0 \tag{D.10}$$

The term which presents difficulty is:

$$\frac{1}{2}\,\text{Trace}\left[\left(\mathbf{L_i}^{\mathbf{T}}\mathbf{L_i}\right)^{-1} \frac{\partial}{\partial \mathbf{z}_{i+1}} \left(\mathbf{L_{i+1}}^{\mathbf{T}}\mathbf{L_{i+1}}\right) \right] \tag{D.11}$$

We shall present the solution to this problem by means of a worked example. Let there be five missing data denoted by z_1, z_2, z_3, z_4, z_5 and let the order of the AR model be three. In addition let

$$\left(\mathbf{L_i}^{\mathbf{T}}\mathbf{L_i}\right)^{-1} = \begin{bmatrix} 4 & 5 & 6 \\ 5 & 7 & -1 \\ 6 & -1 & 2 \end{bmatrix}$$

be an example of a matrix produced during the previous iteration.

Now L_{i+1} has the following structure:

$$[\ L_{i+1}\ ] = \begin{bmatrix} y_3 & y_2 & y_1 \\ \vdots & \vdots & \vdots \\ y_{M-2} & y_{M-3} & y_{M-2} \\ y_{M-1} & y_{M-2} & y_{M-3} \\ z_1 & y_{M-1} & y_{M-2} \\ z_2 & z_1 & y_{M-1} \\ z_3 & z_2 & z_1 \\ z_4 & z_3 & z_2 \\ z_5 & z_4 & z_3 \\ y_K & z_5 & z_4 \\ y_{K+1} & y_K & z_5 \\ y_{K+2} & y_{K+1} & y_K \\ y_{K+3} & y_{K+2} & y_{K+1} \\ \vdots & \vdots & \vdots \\ y_N & y_{N-1} & y_{N-2} \end{bmatrix}$$

For z_1 we can derive

$$\frac{\partial}{\partial z_1}\left(\mathbf{L_{i+1}}^{\mathbf{T}}\ \mathbf{L_{i+1}}\right) = \begin{bmatrix} z_1 & y_{M-1} & y_{M-2} \\ z_2 & z_1 & y_{M-1} \\ z_3 & z_2 & z_1 \end{bmatrix} \tag{D.12}$$

For z_2 we can derive

$$\frac{\partial}{\partial z_2}\left(\mathbf{L_{i+1}}^{\mathbf{T}}\ \mathbf{L_{i+1}}\right) = \begin{bmatrix} z_2 & z_1 & y_{M-1} \\ z_3 & z_2 & z_1 \\ z_4 & z_3 & z_2 \end{bmatrix} \tag{D.13}$$

The pattern is obvious. Differentiating with respect to z_j extracts only rows from $\mathbf{L}_{i+1}$ that contain that element.

The next stage is to premultiply equations D.12 and D.13 by $\left(\mathbf{L_i}^{\mathbf{T}}\ \mathbf{L_i}\right)^{-1}$ and take the trace of the resulting 3×3 matrix. The same can be repeated for the derivatives with respect to z_3, z_4 and z_5.

This gives a term involving the missing data which may be written as follows:

$$\begin{bmatrix} 4+7+2 & 5-1 & 6 & 0 & 0 \\ 5-1 & 4+7+2 & 5-1 & 6 & 0 \\ 6 & 5-1 & 4+7+2 & 5-1 & 6 \\ 0 & 6 & 5-1 & 4+7+2 & 5-1 \\ 0 & 0 & 6 & 5-1 & 4+7+2 \end{bmatrix} \begin{bmatrix} z_1 \\ z_2 \\ z_3 \\ z_4 \\ z_5 \end{bmatrix} + \begin{bmatrix} q_1 \\ q_2 \\ q_3 \\ q_4 \\ q_5 \end{bmatrix}$$

where $\{q_1, q_2, q_3, q_4, q_5\} = \{(5-1)\,y_{M-1} + 6\,y_{M-2},\, 6\,y_{M-1},\, 0,\, 6\,y_K,\, (5-1)\,y_K + 6\,y_{K+1}\}$.

Hence the derivative of the trace is of the form:

$$\frac{1}{2}\,\text{Trace}\left[\left(\mathbf{L_i}^{\mathbf{T}}\,\mathbf{L_i}\right)^{-1}\frac{\partial}{\partial \mathbf{Z}_{i+1}}\left(\mathbf{L_{i+1}}^{\mathbf{T}}\,\mathbf{L_{i+1}}\right)\right] = \mathbf{T}\,\mathbf{z}_{i+1} + \mathbf{q} \tag{D.14}$$

where $\mathbf{T}$ is a symmetric band diagonal Toeplitz matrix. The diagonal elements of $\mathbf{T}$ equal the sum of the elements of the corresponding diagonal of $\left(\mathbf{L_i}^{\mathbf{T}}\,\mathbf{L_i}\right)^{-1}$. The term in $\mathbf{q}$ depends on the observed data and may be calculated efficiently by means of a convolution using the following steps:

1. Augment the observed data with zeroes (i.e. zeroes take the place of the missing data).

2. Convolve the augmented data with a row from $\mathbf{T}$.

3. Extract the elements of the result from the positions occupied by the missing data.

Note that only the observed data near the gap is required when carrying out the convolution.

Appendix E
Issues in Sampling Based Approaches to Integration

Given jointly distributed random variates there are two candidate methods for generating marginal densities. First, one may partition out the variable of interest and estimate the pdf from the samples. The most common approach to estimating a pdf from samples is to use a histogram. Second, one may apply the special case of the dummy variable method derived in section 3.9.3. We describe below how jointly distributed random variates can be used to estimate a marginal density from a conditional density. For the random variates we consider a dummy variable as taking the place of the variable for which the marginal density is required. The conditional density is then averaged out over the remaining components of the parameter vector to obtain an estimate for the marginal density. The advantage of this second approach is that if the conditional density is a continuous curve then the estimate of the marginal density will also be a continuous curve.

We may also use jointly distributed variates to calculate the evidence, providing the corresponding values of the joint density at the sample points are also available. This method, which we term reverse importance sampling, is described in section E.4.

E.1 Marginalizing using the conditional density

In this section, a formal proof will be given to show that the expected value of the conditional density with respect to the joint density is the corresponding marginal density. This result is used in numerical integration to estimate marginal densities.

Consider

$$g(a) = \int_\Theta \int_A f(a \mid \theta) f(\tilde{a}, \theta)\, d\tilde{a}\, d\theta$$

where $f(a \mid \theta)$ and $f(\tilde{a}, \theta)$ denote the conditional and joint posterior densities respectively. By definition,

$$f(\theta) = \int_A f(\tilde{a}, \theta)\, d\tilde{a}$$

giving the expression,

$$g(a) = \int_\Theta f(a \mid \theta) f(\theta)\, d\theta$$

So finally we obtain,

$$g(a) = \int_\Theta f(a, \theta)\, d\theta = f(a)$$

The integral is taken over the entire parameter space. In general, the dimensionality may be quite high so as to make the use of quadrature integration prohibitively expensive. In practice therefore, the integral may be estimated using Monte Carlo methods by the following mixture density:

$$f(a) \approx \frac{1}{M} \sum_{j=1}^{M} f(a \mid \theta_j) \tag{E.1}$$

where the sample variates $\{\tilde{a}_j, \theta_j\}$ are jointly distributed as $f(\tilde{a}, \theta)$. The accuracy of this approximation is of order $O(M^{-\frac{1}{2}})$.

The application of the Gibbs sampler and Hybrid Monte Carlo algorithm to marginalizing posterior densities is now obvious. The output from the Gibbs sampler or Hybrid Monte Carlo algorithm is a sequence of jointly distributed variates. The marginal density is then obtained by averaging the conditional density over these variates using equation E.1.

E.2 Approximating the conditional density

In many cases, the joint density may be evaluated directly but the conditional density will not be available in closed form. We will now describe a completely systematic approach for approximating the conditional density for the purpose of estimating the marginal density.

We may rewrite equation E.1 in the following form:

$$\begin{aligned} f(a) &= \frac{1}{m} \sum_{i=1}^{m} \frac{f(a,\, b_i,\, c_i,\, d_i)}{f(b_i,\, c_i,\, d_i)} \\ &= \frac{1}{m} \sum_{i=1}^{m} w_i\, f(a,\, b_i,\, c_i,\, d_i) \end{aligned} \tag{E.2}$$

where, for the sake of example, we have assumed that parameter space consists of four components $(a,\, b,\, c,\, d)$ with θ replaced by $(b,\, c,\, d)$.

The joint density $f(a,\, b_i,\, c_i,\, d_i)$ may be evaluated at some expense for any value of a. Therefore in order to approximate $f(a)$ some means must be found to evaluate the weights w_i.

We have

$$w_i^{-1} = \int_{\min(a)}^{\max(a)} f(a,\, b_i,\, c_i,\, d_i)\, da \tag{E.3}$$

Typically, the parameters range over the real axis. However, for most probability densities the probability mass is concentrated in one region in space with virtually none of the probability mass in the tails. To a good approximation, even when the integral is improper (i.e. limits are infinite) one can replace $\min(a)$ by the minimum value and $\max(a)$ by the maximum value observed in the jointly distributed variates.

The integral in equation E.3 may be evaluated using Simpson's rule or even by Gauss-Hermite quadrature:

$$w_i^{-1} \approx \sum_{j=1}^{p} \beta_j\, f(a_j,\, b_i,\, c_i,\, d_i) \tag{E.4}$$

where the weights β_j and the abscissae a_j depend on the integration rule.

Suppose Simpson's rule is chosen with $p = 30$. This means that a total of mp samples of the joint density will have to be evaluated. The samples are derived from a grid consisting of two components:

1. m samples of $b_i,\, c_i,\, d_i$ resulting from Monte Carlo random sampling of the joint density.

2. p samples of the parameter of interest a.

This grid therefore has a Monte Carlo component to sample a large dimensional partition of parameter space, and a systematic component (usually uniform) grid for the marginal density taken over the remaining low dimensional partition of parameter space.

There are two stages in computing the marginal density:

1. Estimate the weights w_i.

2. Compute the marginal density using equation E.1.

If the marginal density $f(a_j)$ is calculated at the same points a_j as used in Simpson's rule then there will be no need to compute the joint density during stage 2, since all the necessary computations are carried out during stage 1.

If the samples a_j are uniformly sampled then setting $p = 30$ should give a good representation of the curve. Interpolation may be carried out by low pass filtering the log density and exponentiating the result. If sampling is non-uniform then one may interpolate using splines.

In most cases the computational budget is known in advance. The number of jointly distributed variates used to estimate a marginal density will usually be far less than the total number available. In the case of jointly distributed variates obtained using Markov chains it is better to try to decrease the dependence between samples by choosing samples spaced as far away from each other in time as possible.

E.3 Gibbs sampling from the joint density

From the point of view of the Gibbs sampler, the conditional density is proportional to the joint density.

$$p(a_i \mid a_j \; i \neq j) \qquad \propto \qquad p(a_i, \, a_j \; i \neq j) \tag{E.5}$$

The constant of proportionality $[p(a_j \; i \neq j)]^{-1}$ does not actually matter from the point of view of drawing random variates. In other words, drawing random variates from the joint density according to the Gibbsian scheme is equivalent to drawing from the conditional densities.

The only difficulty at this point is how to draw random variates from a cross section through the joint density. Often, such as in the audio restoration example in chapter 6 this can be done directly. However there are many examples where local Metropolis steps may have to be applied.

E.4 Reverse importance sampling

In this section, we describe a method for estimating the evidence of a model given both random samples from the joint density and the values of the joint density at these sample points.

The method is based on importance sampling but differs in that the samples are drawn from the joint density. These are used to integrate a function for which the normalization constant is known, thereby obtaining the value of the unknown normalization constant (the evidence) relative to a known normalization constant for the integrand.

Consider the integral

$$\int_{\mathbf{X}} g(\mathbf{x}) \, d\mathbf{x} = I \tag{E.6}$$

where we have chosen $g(\mathbf{x})$ such that $I = 1$.

The evidence E is defined by

$$E = \int_{\mathbf{X}} f(\mathbf{x})\, d\mathbf{x} \tag{E.7}$$

Let us rewrite equation E.6 in the form

$$I = \int_{\mathbf{X}} \frac{g(\mathbf{x})}{f(\mathbf{x})} f(\mathbf{x})\, d\mathbf{x} \tag{E.8}$$

Dividing both sides of equation E.8 by E to normalize $f(\mathbf{x})$ we obtain

$$\frac{I}{E} = \int_{\mathbf{X}} \frac{g(\mathbf{x})}{f(\mathbf{x})} \frac{f(\mathbf{x})}{E}\, d\mathbf{x} \tag{E.9}$$

The right hand side of equation E.9 may be readily evaluated using randomly distributed sample variates from $f(\mathbf{x})$ as:

$$\frac{I}{E} = E_g\left(\frac{g(\mathbf{x})}{f(\mathbf{x})}\right) \approx \frac{1}{N}\sum_{i=1}^{N} \frac{g(x_i)}{f(x_i)} \tag{E.10}$$

It follows that since I is known therefore E can be estimated.

The Monte Carlo integral in equation E.10 is different from equation 3.68. On one hand, equation 3.68 estimates evidence by comparing $f(\mathbf{x})$ with a normalized $g(\mathbf{x})$ by sampling from $g(\mathbf{x})$. On the other hand, in equation E.10 the evidence is computed by comparing a normalized $g(\mathbf{x})$ with $f(\mathbf{x})$ by sampling from $f(\mathbf{x})$. As we have seen, equation 3.68 may be implemented using random numbers generated using non-Markov chain techniques, whereas to implement equation E.10 one may have to sample directly from the integrand using a Markov chain technique.

As always when carrying out importance sampling, it is desirable that the sampling density matches the integrand as closely as possible. Because $g(\mathbf{x})$ is the numerator of equation E.10 then for numerical stability it is desirable that it have *thinner* tails than $f(\mathbf{x})$. Using the principle of asymptotic normality as a guide leads one to choose a Gaussian $g(\mathbf{x})$.

The mean and the covariance matrix of $g(\mathbf{x})$ may be estimated from N random samples $\mathbf{X}_i$ drawn from $f(\mathbf{x})$ by means of the following expression

$$\hat{\mu} = \frac{1}{N}\sum_{i=1}^{N} \mathbf{X}_i \tag{E.11}$$

$$\hat{\mathbf{C}} = \frac{1}{N-1}\sum_{i=1}^{N} (\mathbf{X}_i - \hat{\mu})(\mathbf{X}_i - \hat{\mu})^{\mathbf{T}} \tag{E.12}$$

Using estimates of the mode and the covariance matrix from the same variates used to compute the integral obviously introduces bias into the final estimate of the evidence. This should not worry us unduly; Hammersley and Handscomb [47] use sample estimates of the covariance matrix (albeit in a slightly different context) for Monte Carlo integration with excellent results.

Appendix F
Detailed Balance

It is very informative to show that the Gibbs sampler, Metropolis algorithm and the Hybrid Monte Carlo algorithm all obey the condition of detailed balance. If detailed balance holds then this implies that the joint density is an invariant distribution.

In all that follows the notation $\text{Trans}(\mathbf{X}, \mathbf{X}')$ denotes the probability of a transition from $\mathbf{X}$ to $\mathbf{X}'$.

F.1 Detailed balance in the Gibbs sampler

Let us consider a transition from a point $(\mathbf{B}, \mathbf{A})$ to another point $(\mathbf{C}, \mathbf{A})$ in state space. If we assume that the Gibbs sampler has attained equilibrium then the probability of the transition is:

$$\text{Trans}\left[(\mathbf{B}, \mathbf{A}), (\mathbf{C}, \mathbf{A})\right] = p(\mathbf{B}, \mathbf{A})\, p(\mathbf{C} \mid \mathbf{A}) \tag{F.1}$$

The probability of a transition in the reverse direction is:

$$\text{Trans}\left[(\mathbf{C}, \mathbf{A}), (\mathbf{B}, \mathbf{A})\right] = p(\mathbf{C}, \mathbf{A})\, p(\mathbf{B} \mid \mathbf{A}) \tag{F.2}$$

Now

$$p(\mathbf{B}, \mathbf{A})\, p(\mathbf{C} \mid \mathbf{A}) = \frac{p(\mathbf{B}, \mathbf{A})\, p(\mathbf{C}, \mathbf{A})}{p(\mathbf{A})} = p(\mathbf{B} \mid \mathbf{A})\, p(\mathbf{C}, \mathbf{A}) \tag{F.3}$$

Therefore expression F.1 is equal to expression F.2 and detailed balance holds.

F.2 Detailed balance in the Metropolis-Hastings algorithm

Let us consider a transition from a point $\mathbf{X}$ in parameter space to a point $\mathbf{X}'$. The proposal density is $T(\mathbf{X}, \mathbf{X}')$. If we assume that the Markov chain has attained equilibrium then the probability of transition is:

$$\text{Trans}\,[\mathbf{X}, \mathbf{X}'] = p(\mathbf{X})\,T(\mathbf{X}, \mathbf{X}')\,\min\left[1, \frac{p(\mathbf{X}')T(\mathbf{X}', \mathbf{X})}{p(\mathbf{X})T(\mathbf{X}, \mathbf{X}')}\right] \tag{F.4}$$

Since probabilities are non-negative we may multiply through to obtain:

$$\text{Trans}\,[\mathbf{X}, \mathbf{X}'] = \min\left[p(\mathbf{X})\,T(\mathbf{X}, \mathbf{X}'), p(\mathbf{X}')T(\mathbf{X}', \mathbf{X})\right] \tag{F.5}$$

The probability of a transition from point $\mathbf{X}'$ to $\mathbf{X}$ is:

$$\text{Trans}\,[\mathbf{X}', \mathbf{X}] = p(\mathbf{X}')\,T(\mathbf{X}', \mathbf{X})\,\min\left[1, \frac{p(\mathbf{X})T(\mathbf{X}, \mathbf{X}')}{p(\mathbf{X}')T(\mathbf{X}', \mathbf{X})}\right] \tag{F.6}$$

Multiplying through gives:

$$\text{Trans}\,[\mathbf{X}', \mathbf{X}] = \min\left[p(\mathbf{X}')\,T(\mathbf{X}', \mathbf{X}), p(\mathbf{X})T(\mathbf{X}, \mathbf{X}')\right] \tag{F.7}$$

Since it is always true that $\min(A, B) = \min(B, A)$ then expression F.5 and expression F.7 are identical and hence detailed balance holds.

F.3 Detailed balance in the Hybrid Monte Carlo algorithm

Let us consider a transition from a point $\mathbf{X}$ in phase space with Hamiltonian H to another point $\mathbf{X}'$ with Hamiltonian H' in phase space which is obtained by tracing the dynamics forward in time. If we assume that the Markov chain has attained equilibrium then the probability of transition is:

$$\begin{aligned}\text{Trans}\,[\mathbf{X}, \mathbf{X}'] &= \exp[-H]\,\frac{1}{2}\,\min\left[1, \exp(-(H' - H))\right] \\ &= \frac{1}{2}\,\min\left[\exp(-H), \exp(-H')\right] \end{aligned} \tag{F.8}$$

The factor of 1/2 indicates the probability that the dynamics will be traced forward in time. The probability of a transition from point $\mathbf{X}'$ to $\mathbf{X}$ is given by:

$$\begin{aligned}\text{Trans}\,[\mathbf{X}', \mathbf{X}] &= \exp[-H']\,\frac{1}{2}\,\min\left[1, \exp(-(H - H'))\right] \\ &= \frac{1}{2}\,\min\left[\exp(-H'), \exp(-H)\right] \end{aligned} \tag{F.9}$$

The factor of $1/2$ indicates the probability that the dynamics will be traced backward in time. As before in the case of the Metropolis algorithm, since it is always true that $\min(A, B) = \min(B, A)$ then expression F.8 is equal to expression F.9 and therefore detailed balance holds.

F.4 Remarks

The above proofs show that the joint distribution is an invariant distribution of the three kinds of Markov chains used in the dissertation. This result does not completely prove that the chains are useful for sampling from the joint density. For that to be the case:

1. The invariant distribution must be unique.
2. The chain must be ergodic.

We have shown that the second result holds for discrete regular Markov chains.

Geman and Geman [36] give complete proofs that the Gibbs sampler is ergodic and samples from the joint density uniquely. The Hybrid Monte Carlo algorithm will sample the joint density uniquely because of Liouville's theorem and the conditions of detailed balance that we give above. The general Metropolis-Hastings algorithm is more difficult to analyse.

References

[1] H. Akaike, *Information theory and the extension of the maximum likelihood principle*, in Second International Symposium on Information Theory, B. N. Petrov and F. Csaki, eds., Budapest, 1973, pp. 267–281.

[2] M. Al-Chalabi, *When least-squares least*, Geophysical Prospecting, 40 (1992), pp. 359–378.

[3] B. J. Alder and T. E. Wainwright, *Studies in molecular dynamics I: General method*, Journal of Chemical Physics, 31 (1959), pp. 459–466.

[4] H. C. Andersen, *Molecular dynamics simulations at constant pressure and/or temperature*, Journal of Chemical Physics, 72 (1980), pp. 2384–2393.

[5] D. H. Anderson, *Compartmental Modelling and Tracer Kinetics, Lecture Notes in Biomathematics*, Springer-Verlag, 1983.

[6] G. Arfken, *Mathematical Methods for Physicists*, Academic Press, 1985.

[7] P. W. Atkins, *Physical Chemistry*, 2nd ed., Oxford University Press, 1982.

[8] O. Barndorff-Nielsen and D. R. Cox, *Edgeworth and saddlepoint approximations with statistical applications*, Journal of the Royal Statistical Society B, 41 (1979), pp. 279–312.

[9] P. A. Bash, U. C. Singh, R. Langridge, and P. A. Kollman, *Free energy calculations by computer simulation*, Science, 236 (1987), pp. 564–568.

[10] C. H. Bennett, *Efficient estimation of free energy differences from Monte Carlo data*, Journal of Computational Physics, 22 (1976), pp. 245–268.

[11] J. M. BERNARDO AND A. F. M. SMITH, *Bayesian Theory*, John Wiley and Sons Ltd., 1994.

[12] P. R. BEVINGTON, *Data Reduction and Error Analysis for the Physical Sciences*, McGraw-Hill, 1969.

[13] G. L. BRETTHORST, *Bayesian Spectrum Analysis and Parameter Estimation*, Springer-Verlag, 1989.

[14] ———, *Bayesian analysis I: Parameter estimation using quadrature NMR models*, Journal of Magnetic Resonance, 88 (1990), pp. 533–551.

[15] ———, *Bayesian analysis II: signal detection and model selection*, Journal of Magnetic Resonance, 88 (1990), pp. 552–570.

[16] ———, *Bayesian analysis III: Applications to NMR signal detection, model selection, and parameter estimation*, Journal of Magnetic Resonance, 88 (1990), pp. 572–595.

[17] D. BURTON AND W. J. FITZGERALD, *Bayesian parameter estimation: Further results*, tech. report, Marconi Maritime Research Laboratory, Cambridge, England, 1989.

[18] ———, *Bayesian spectrum analysis and parameter estimation*, tech. report, Marconi Maritime Research Laboratory, Cambridge, England, 1989.

[19] ———, *Model selection by Bayesian analysis*, tech. report, Marconi Maritime Research Laboratory, Cambridge, England, 1989.

[20] D. BURTON, G. J. MOORE, AND W. J. FITZGERALD, *Bayesian model selection and parameter estimation applied to sea floor pressure data*, in Maximum Entropy and Bayesian Methods, P. F. Fougere, ed., Kluwer Academic Publishers, Netherlands, 1990, pp. 185–194.

[21] B. B. CARLIN, A. E. GELFAND, AND A. F. M. SMITH, *Hierarchical Bayesian analysis of changepoint problems*, Applied Statistics, 41 (1992), pp. 389–405.

[22] C. CHATFIELD, *The Analysis of Time Series*, London, Chapman and Hall, 1975.

[23] M. H. CHEN, *Importance weighted marginal Bayesian posterior density estimation*, tech. report, Department of Statistics, Purdue University, 1992.

[24] R. T. COX, *Probability, frequency and reasonable expectation*, American Journal of Physics, 14 (1946), pp. 1–13.

[25] D. G. CRIGHTON, A. P. DOWLING, J. E. FFOWCS-WILLIAMS, AND F. G. LEPPINGTON, *Aero and hydro acoustics: An advanced short course*, tech. report, NASA Ames Research Center, 1984.

[26] H. E. DANIELS, *The approximate distribution of serial correlation coefficients*, Biometrika, 43 (1956), pp. 169–185.

[27] P. J. DAVIS AND I. POLONSKY, *Numerical interpolation, differentiation and integration*, in Handbook of Mathematical Functions, M. Abramowitz and A. Stegun, eds., The National Bureau of Standards (U. S. A.), 1965, pp. 875–924.

[28] P. J. DAVIS AND P. RABINOWITZ, *Methods of Numerical Integration*, Academic Press, 1975.

[29] A. P. DEMPSTER, N. LAIRD, AND D. B. RUBIN, *Maximum likelihood from incomplete data via the EM algorithm*, Journal of the Royal Statistical Society, B (1977), pp. 1–38.

[30] L. DEVROYE, *Non-uniform Random Variate Generation*, Springer-Verlag, 1986.

[31] S. DUANE, A. D. KENNEDY, B. J. PENDLETON, AND D. ROWETH, *Hybrid Monte Carlo*, Physics Letters B, 195 (1987), pp. 216–222.

[32] W. J. FITZGERALD, *Bayesian inference and signal processing*, tech. report, Cambridge University Engineering Department, 1991. CUED/F-INFENG/TR80.

[33] W. J. FITZGERALD AND M. NIRANJAN, *Speech processing using Bayesian inference*, in Bayesian Statistics and Maximum Entropy Conference, Paris, August 1992, Kluwer Academic Press.

[34] R. FLETCHER, *Practical Methods of Optimization*, 2nd ed., John Wiley and sons, 1987.

[35] A. E. GELFAND AND A. F. M. SMITH, *Sampling based approaches to calculating marginal densities*, Journal of the American Statistical Association, 85 (1990), pp. 398–409.

[36] S. GEMAN AND D. GEMAN, *Stochastic relaxation, Gibbs distributions, and the Bayesian restoration of images*, IEEE Trans. on Pattern Analysis and Machine Intelligence, PAMI-6 (1984), pp. 721–741.

[37] W. GERSCH AND G. KITIGAWA, *Smoothness priors in time series*, in Bayesian Analysis of Time Series and Dynamic Models, New York; Marcel Dekker, 1988, ch. 16, pp. 431–476.

[38] J. GEWEKE, *Antithetic acceleration of Monte Carlo integration in Bayesian inference*, Journal of Econometrics, 88 (1988), pp. 73–89.

[39] W. R. GILKS AND P. WILD, *Adaptive rejection sampling for Gibbs sampling*, Applied Statistics, (1992), pp. 337–348.

[40] S. J. GODSILL, *The Restoration of Degraded Audio Signals*, PhD Thesis, Cambridge University Engineering Department, England, 1993.

[41] S. J. GODSILL AND P. J. W. RAYNER, *Robust reconstruction and analysis of autoregressive signals in impulsive noise using the Gibbs sampler*, tech. report, Cambridge University Engineering Department, England, 1996. CUED/F-INFENG/TR223.

[42] D. E. Goldberg, *Genetic Algorithms in Search, Optimization and Machine Learning*, Addison-Wesley, 1989.

[43] G. H. Golub and C. F. Van Loan, *Matrix Computations*, The John Hopkins University Press, 1990.

[44] K. Gordon and A. F. M. Smith, *Modelling and monitoring discontinuous changes in time series*, tech. report, Department of Mathematics, University of Nottingham, England, 1992.

[45] P. J. Green, *Reversible jump Markov chain Monte Carlo computation and Bayesian model determination*, in The Workshop on Model Criticism in Highly Structured Stochastic Systems, Wiesbaden, Germany, September 1994.

[46] U. Grenander and M. I. Miller, *Representation of knowledge in complex systems*, Journal of the Royal Statistical Society-B, 56 (1994), pp. 549–603.

[47] J. C. Hammersley and D. C. Handscomb, *Monte Carlo Methods*, Methuen's statistical monographs, Wiley, 1964.

[48] W. K. Hastings, *Monte Carlo sampling methods using Markov chains and their applications*, Biometrika, 57 (1970), pp. 97–109.

[49] S. E. Hills and A. F. M. Smith, *Diagnostic plots for improved parameterization in Bayesian inference*, Biometrika, 80 (1991), pp. 61–74.

[50] ——, *Parameterization issues in Bayesian inference*, in Bayesian Statistics 4, J. Bernardo, ed., Oxford University Press, 1992.

[51] R. Hooke and T. A. Jeeves, *Direct search solution of numerical and statistical problems*, J. Assoc. Comp. Mach., (1962), pp. 212–229.

[52] J. E. Hudson, *Basic mathematics of adaptive filters*, in Mathematics of Adaptive Filtering, University of Warwick, December 1988.

[53] L. Ingber and B. Rosen, *Genetic algorithms and very fast simulated reannealing*, Mathematical and Computer Modelling, 16 (1992), pp. 87–100.

[54] J. A. Jacquez, *Compartmental Analysis in Biology and Medicine*, Elsevier, 1972.

[55] E. T. Jaynes, *Bayesian methods: General background*, in The Fourth Annual Workshop on Bayesian/Maximum Entropy Methods in Geophysical Inverse Problems, University of Calgary, August 1984.

[56] ——, *Bayesian spectrum and chirp analysis*, in Maximum-Entropy and Bayesian Spectral Analysis and Estimation Problems, C. R. Smith and G. J. Erickson, eds., New York: D. Reidel Publishing Company, 1987.

[57] ——, in Papers on Probability, Statistics and Statistical Physics, a Reprint Collection, R. D. Rosenkrantz, ed., Kluwer, 1989.

[58] H. Jeffreys, *Theory of Probability*, Oxford University Press, 1939.

[59] M. Kalos and P. Whitlock, *Monte Carlo Methods*, Wiley, 1986.

[60] R. E. Kass, L. Tierney, and J. B. Kadane, *Laplace's method in Bayesian analysis*, Contemporary Mathematics, 115 (1991), pp. 89–99.

[61] A. D. Kennedy, *The theory of hybrid stochastic algorithms*, in Probabilistic Methods in Quantum Field Theory and Quantum Gravity, New York: Plenum Press, 1990.

[62] M. Kheradmandnia, *Aspects of Bayesian threshold autoregressive modelling*, PhD Thesis, University of Kent, England, 1991.

[63] S. Kirkpatrick, *Optimization by simulated annealing – qualitative studies*, Journal of Statistical Physics, 34 (1984), pp. 975–986.

[64] S. Kirkpatrick, C. D. Gellatt, and M. P. Vecchi, *Optimization by simulated annealing*, Science, 220 (1983), pp. 671–680.

[65] T. Kloek and H. K. Van Dijk, *Bayesian estimates of equation system parameters: An application of integration by Monte Carlo*, Econometrica, 46 (1978), pp. 1–19.

[66] E. Kreyzsig, *Advanced Engineering Mathematics*, Wiley, 1983.

[67] C. Lanczos, *Applied Analysis*, Pitman, 1957.

[68] J. Lasenby and W. J. Fitzgerald, *A Bayesian approach to high-resolution beamforming*, IEE Proceedings-F, 138 (1991), pp. 539–544.

[69] ——, *Entropic Volterra classifier (EVC) for use in data classification*, Electronics Letters, 30 (1994), pp. 53–54.

[70] P. M. Lee, *Bayesian statistics: An introduction*, Oxford University Press, 1989.

[71] T. Leonard, S. J. Hsu, and K. Tsui, *Bayesian marginal inference*, J. Am. Statist. Ass., 84 (1989), pp. 1051–1058.

[72] G. P. Lepage, Journal of Computational Physics, 27 (1978), pp. 192–203.

[73] D. V. Lindley, *Approximate Bayesian methods*, in Bayesian Statistics, J. M. Bernardo, ed., Valencia; Valencia University Press, 1980, pp. 223–245.

[74] D. MacKay, *Bayesian Methods for Adaptive Models*, PhD Thesis, California Institute of Technology, December 1991.

[75] R. J. Marks, G. L. Wise, D. G. Haldeman, and J. L. Whited, *Detection in Laplace noise*, IEEE Transactions on Aerospace and Electronic Systems, AES-14 (1978), pp. 866–871.

[76] R. E. McCulloch and R. S. Tsay, *Bayesian analysis of autoregressive time series via the Gibbs sampler*, Journal of Time Series Analysis, 15 (1994), pp. 235–250.

[77] N. METROPOLIS, A. W. ROSENBLUTH, M. N. ROSENBLUTH, A. H. TELLER, AND E. TELLER, *Equation of state calculations by fast computing machines*, Journal of Chemical Physics, 21 (1953), pp. 1087–1092.

[78] R. D. MORRIS AND W. J. FITZGERALD, *Line scratch detection and removal using reversible jump Markov chain Monte Carlo methods*, in IEEE International Conference on Acoustics Speech and Signal Processing, Atlanta, GA USA, 1996. To be published.

[79] F. MOSTELLER AND D. L. WALLACE, *Inference and disputed authorship: The federalist*, Addison-Wesley, 1964.

[80] J. NAYLOR, *Econometric illustrations of novel numerical integration strategies for Bayesian inference*, Journal of Econometrics, 38 (1988), pp. 103–125.

[81] J. C. NAYLOR AND A. F. M. SMITH, *Applications of a method for the efficient computation of posterior distributions*, Applied Statistics, 31 (1982), pp. 214–225.

[82] R. NEAL, *Probabilistic inference using Markov chain Monte Carlo methods*, Tech. Report CRG-TR-93-1, Department of Computer Science, University of Toronto, 1993.

[83] J. A. NELDER AND R. MEAD, *A simplex method for function minimization*, Computer Journal, 7 (1965), pp. 308–313.

[84] E. NUMMELIN, *General irreducible Markov chains and non-negative operators*, Cambridge University Press, 1984.

[85] J. J. K. Ó RUANAIDH AND W. J. FITZGERALD, *The identification of discontinuities in time series using the Bayesian piecewise linear model*, tech. report, Department of Engineering, University of Cambridge, 1993. CUED/F-INFENG/TR134.

[86] J. J. K. Ó RUANAIDH, W. J. FITZGERALD, AND K. J. POPE, *Recursive Bayesian location of a discontinuity in time series*, in International Conference on Acoustics, Speech and Signal Processing, Adelaide, Australia, 1994.

[87] A. V. OPPENHEIM AND R. W. SCHAFER, *Discrete-time Signal Processing*, Prentice Hall, 1989.

[88] R. H. J. M. OTTEN AND L. P. P. P. VAN GINNEKEN, *Floorplan design using simulated annealing*, in IEEE International Conference on Computer Aided Design (ICCAD-84), 1984.

[89] A. PAPOULIS, *Probability, Random Variables and Stochastic Processes*, 2nd ed., McGraw-Hill, 1984.

[90] P. Z. PEEBLES, *Probability, Random variables and Random Signal Principles*, 3rd ed., McGraw-Hill, 1993.

[91] P. H. PESKUN, *Optimum Monte Carlo sampling using Markov chains*, Biometrika, 60 (1973), pp. 607–612.

[92] D. B. Phillips and A. F. M. Smith, *Bayesian model comparison via jump diffusions*, tech. report, Imperial College of Science, Technology and Medicine, September 1994.

[93] E. Polak, *Computational Methods in Optimization*, Academic Press, 1971.

[94] K. J. Pope, *Time Series Analysis*, PhD Thesis, Cambridge University Engineering Department, England, 1993.

[95] W. H. Press, S. A. Teukolsky, W. T. Vetterling, and B. P. Flannery, *Numerical Recipes in C*, 2nd ed., Cambridge University Press, 1992.

[96] R. Prony, J. l'Ecole Polytechnique, (1795), pp. 24–76.

[97] A. P. Quinn, *Bayesian Point Inference in Signal Processing*, PhD Thesis, Cambridge University Engineering Department, England, September 1992.

[98] A. P. Quinn and M. D. MacLeod, *The distinction between joint and marginal estimators*, in Fourth Valencia International meeting on Bayesian Statistics, 15-20 April 1991.

[99] J. J. Rajan, *Time Series Classification*, PhD Thesis, Cambridge University Engineering Department, England, October 1994.

[100] J. J. Rajan and P. J. W. Rayner, *Parameter estimation of time-varying autoregressive models using the Gibbs sampler*, Electronics Letters, 31 (1995), pp. 1035–1036.

[101] J. J. Rajan, P. J. W. Rayner, and S. J. Godsill, *A Bayesian approach to parameter estimation and interpolation of time-varying autoregressive processes using the Gibbs sampler.* Submitted to the IEEE Transactions on Signal Processing, May 1995.

[102] C. R. Rao, *Linear Statistical Inference and Applications*, Wiley, 1973.

[103] P. J. W. Rayner and S. J. Godsill, *The detection and correction of artefacts in archived gramophone recordings*, in Proc. IEEE ASSP Workshop on Applications of Signal Processing to Audio and Acoustics, Mohonk, NY State, 1991.

[104] B. D. Ripley, *Stochastic Simulation*, John Wiley, 1987.

[105] J. Rissanen, *Modeling by shortest data description*, Automatica, 14 (1978), pp. 465–471.

[106] C. Ritter and M. A. Tanner, *The Gibbs stopper and the griddy Gibbs sampler*, J. Am. Statist. Ass., 87 (1992), pp. 861–867.

[107] J. J. K. Rooney and W. J. Fitzgerald, *Bayesian model based parameter estimation and model selection in impulsive noise environments*, in International Conference on Acoustics, Speech and Signal Processing, Minneapolis, U. S. A., 1993.

[108] R. Y. Rubenstein, *Simulation and the Monte Carlo Method*, Wiley, 1981.

[109] D. B. RUBIN, *Using the SIR algorithm to simulate posterior distributions*, in Bayesian Statistics 3, J. Bernardo et al., eds., Oxford University Press, 1988.

[110] P. SALAMON, J. D. NULTON, J. PEDERSON, G. RUPPEINER, AND L. LIAO, *Simulated annealing with constant thermodynamic speed*, Computer Physics Communications, 49 (1988), pp. 423–428.

[111] J. A. SCALES, M. L. SMITH, AND T. L. FISCHER, *Global optimization methods for multimodal inverse problems*, Journal of Computational Physics, 103 (1992), pp. 258–268.

[112] J. SCHROEDER, R. YARLAGADDA, AND J. HERSHEY, *L_p normed minimization with applications to linear predictive modeling for sinusoidal frequency estimation*, Signal Processing, 24 (1991), pp. 193–216.

[113] G. SCHWARZ, *Estimating the dimension of a model*, The Annals of Statistics, 6 (1978), p. 461.

[114] S. SHANMUGAN AND A. BREIPOHL, *Random Signals: Detection, Estimation and Data Analysis*, John Wiley and Sons, 1988.

[115] J. E. H. SHAW, *A quasirandom approach to integration in Bayesian statistics*, The Annals of Statistics, 16 (1988), pp. 895–914.

[116] S. SIBISI AND J. SKILLING, *Bayesian density estimation*. University of Cambridge, Department of Physics; submitted to Journal of Royal Statistical Society, 25th March 1994.

[117] J. SKILLING, *Classic maximum entropy*, in Maximum Entropy and Bayesian Methods, Kluwer Academic Publishers, 1989, ch. Classic Maximum Entropy, pp. 45–52.

[118] A. F. M. SMITH, *A Bayesian approach to inference about a changepoint in a sequence of random variables*, in Biometrika, 1975, pp. 407–416.

[119] ——, *Bayesian computational methods*, Philsophical Transactions of the Royal Society of London, A 337 (1991), pp. 369–386.

[120] A. F. M. SMITH, A. M. SKENE, J. E. H. SHAW, J. C. NAYLOR, AND M. DRANSFIELD, *The implementation of the Bayesian paradigm*, Communications in Statistics, A14 (1985), pp. 1079–1102.

[121] D. A. STEPHENS, *Bayesian retrospective multiple changepoint identication*, tech. report, Department of Mathematics, Imperial College, London, England, 1992.

[122] ——, *Bayesian retrospective multiple-changepoint identification*, Applied Statistics, 43 (1994), pp. 1159–178.

[123] D. A. STEPHENS AND A. F. M. SMITH, *Bayesian edge detection in images via changepoint methods*, tech. report, Department of Mathematics, Imperial College, London, England, 1992.

[124] A. STROUD, *Approximate Computation of Multiple Integrals*, Prentice-Hall, 1971.

[125] H. SZU AND R. HARTLEY, *Fast simulated annealing*, Physics Letters A., 122 (1987), pp. 157–162.

[126] M. A. TANNER, *Tools for Statistical Inference*, Springer-Verlag, 1993.

[127] M. A. TANNER AND W. H. WONG, *The calculation of posterior distributions by data augmentation*, J. Am. Statist. Ass., 82 (1987), pp. 528–539.

[128] C. W. THERRIEN, *Discrete Random Signals and Statistical Signal Processing*, Prentice Hall International, 1992.

[129] B. TIDOR, *Simulated annealing on free energy surfaces by a combined molecular dynamics and Monte Carlo approach*, Journal of Physical Chemistry, 97 (1993), pp. 1067–1073.

[130] L. TIERNEY, *Markov chains for exploring posterior distributions*, tech. report, School of Statistics, University of Minnesota, 1991.

[131] L. TIERNEY AND J. B. KADANE, *Accurate approximation for posterior moments and marginal densities*, J. Am. Statist. Ass., 81 (1986), pp. 82–86.

[132] L. TIERNEY, R. E. KASS, AND J. B. KADANE, *Approximate marginal densities of non-linear functions*, Tech. Report 513, University of Minnesota, Department of Statistics, 1988.

[133] ——, *Approximate marginal densities of nonlinear functions*, Biometrika, 76 (1989), pp. 425–434.

[134] ——, *Fully exponential Laplace approximations to expectations and variances of nonpositive functions*, J. Am. Statist. Ass., 84 (1989), pp. 710–716.

[135] G. M. TORRIE AND J. P. VALLEAU, *Nonphysical sampling distributions in Monte Carlo free energy estimation: Umbrella sampling*, Journal of Computational Physics, 23 (1977), pp. 187–199.

[136] S. VASEGHI, *Algorithms for the Restoration of Archived Gramophone Recordings*, PhD Thesis, Cambridge University Engineering Department, England, 1988.

[137] R. VELDHUIS, *Restoration of Lost Samples in Digital Signals*, Prentice Hall International, 1990.

[138] A. M. WALKER, *On the asymptotic behaviour of posterior distributions*, Journal of the Royal Statistical Society, 31 (1969), pp. 80–88.

[139] R. L. WOLPERT, *Monte Carlo integration in Bayesian statistical analysis.* Unpublished, 1992.

[140] J. A. YATES, *Changepoint Detection using Bayesian Inference*, Master's Thesis, Cambridge University Engineering Department, England, 1994.

[141] S. L. ZEGER AND M. R. KARIM, *Generalized linear models with random effects: A Gibbs sampling approach*, J. Am. Statist. Ass., 86 (1991), pp. 79–86.

Index